PRAISE

"A deeply researched, thoroughly researched book on the nature and character of warfare and its implications in the modern world. This is a must-read for practitioners, theorists and students of the art of war. Highly recommended."
— **LCol Professor David Kilcullen, PhD**, author of *The Accidental Guerrilla* and *The Dragons and the Snakes: How the Rest Learned to Fight the West*

"Oliviero's Strategia is a welcome antidote to the malaise that currently afflicts Western Professional Military Education (PME) schools that try to train and educate strategic thinkers via 50-minute PowerPoint presentations and links to YouTube videos. Oliviero's "primer" on strategy and doctrine vigorously leads its readers through centuries of western military thought, and, as it should, it demands much from them. PME students should not shy from the challenge, however: with a closely read and marked-up copy of Strategia in hand, they will know where to look within the canon, and what to look for in it. Oliviero gives PME students a much-needed guide to becoming well-informed and deeply versed strategists, which we can only hope will free them from the curse of parroting bromides they learned at staff college as a substitute for serious and rigorous thought."
— **Dr. John Grenier**, author of *The First Way of War: American War Making on the Frontier, 1607-1814*

"The world order has once again fundamentally changed. NATO Secretary General Jens Stoltenberg described the new normal as assertiveness turning to aggression and international law and norms being ignored by revisionist states. More than ever, senior military leaders, strategists and national leaders need to reflect deeply on the true nature of armed conflict. Colonel (retired) Chuck Oliviero provides through his book Strategia an enlightened perspective on military thought and theory, one that will stimulate a more profound reflection and will guide both military professionals and those interested in strategy and military history in their professional development. This primer is a must read for all the students of our respective War Colleges. Chuck has once again proven himself as an eminent military scholar and practitioner and a gifted communicator who can expose his audience to one of the most complex topics, that of military art and science in the context of social history. Professionals must never stop learning and exploring; Chuck's Strategia is a great guide for that."
— **Lieutenant-General (ret'd) J.M. Lanthier**, Former Vice Chief of the Defence Staff of the Canadian Armed Forces

"These days, we fight wars but we do not win them. Worse, we don't really understand what we've done to ourselves and to our enemies. In Strategia, Colonel Charles S. Oliviero, Canadian Army, Retired, explains why we're getting it wrong. Victory in war stems from thinking it through. And thinking it through starts with thinking, no easy matter for busy military leaders, harried political chiefs, or distracted citizens. In a witty and perceptive narrative grounded in military history and the classic works of military thought, Chuck Oliviero shows us the way to go at it. The colonel knows the deal."

— **Daniel P. Bolger, Lieutenant General, U.S. Army, Retired, former combat commander in Iraq and Afghanistan, and author of** *Why We Lost: A General's Inside Account of the Iraq and Afghanistan Wars*

"Strategia is more than a survey of the great theorists of warfare, although it is certainly that. Going far beyond a simple summation, Charles Oliviero offers a coherent critique of modern military ideas, and especially the persistent tendency to generalize from a technological advantage into a supposedly superior approach to war as a whole. Not only soldiers and academics but the general public can profit from a close study of this innovative approach to the nature of armed conflict."

— **Jonathan M. House, Professor Emeritus, U.S. Army Command and General Staff College**

"I have known Chuck all my adult life serving together as brother officers in our regiment and at various military colleges at home and abroad teaching senior military officers the art and science of warfare. He won his tactical spurs in the Regiment rising to command. Throughout he has pursued with passion professional military education to become one of the nation's great soldier-scholars. After three decades of globalization, the chimera of the peace dividend and security provided by the then Leviathan, the West has become complacent in the belief that never again will Europe be engulfed in conflict. Now more than ever, given the emergence of Cold War 2.0, all those searching for an understanding of national security, strategic thought and the application of military art should study this important text."

— **Colonel (ret'd) Chris Corrigan, CD, MA. Former Director of National Security Studies, Canadian Forces College**

STRATEGIA

A PRIMER ON THEORY AND STRATEGY FOR STUDENTS OF WAR

STRATEGIA

A PRIMER ON THEORY AND STRATEGY FOR STUDENTS OF WAR

by

Colonel Charles S. Oliviero

Copyright 2022 Charles S. Oliviero

All rights reserved. No part of this publication may be reproduced or transmitted in any form or by any means, electronically or mechanically including photocopying, recording or any information storage or retrieval system, without prior permission in writing from the author.

Library and Archives Canada Cataloguing in Publication
Oliviero, Charles S. author
Strategia / Charles S. Oliviero

Issued in print and electronic formats.
ISBN: 978-1-990644-24-5 (soft cover)
ISBN: 978-1-990644-25-2 (e-book)

Editor: Phil Halton
Cover and interior design: Pablo Javier Herrera

Cover Photo credit: Corinthian helmet by Keith Binns.

The author would also like to thank the Commanding Officer and gratefully acknowledge the use of the Regimental crest of The Queen's York Rangers (1st American Regiment) (RCAC).

Double Dagger Books Ltd
Toronto, Ontario, Canada
www.doubledagger.ca

DEDICATION

This volume is dedicated to the members of The Queen's York Rangers (1st American Regiment) (RCAC), Canada's oldest cavalry regiment. My decades-long association with the Ranger family began in the late 1970s when as a young Battle Captain, I was assigned to assist them in their annual summer training at Canadian Forces Base Petawawa. It came full circle in 2016 when I was honoured to be appointed by the Defence Minister as the Regiment's Honorary Lieutenant Colonel, thereby allowing me to close off my life of military service in the uniform of a cavalryman.

The Regiment's proud tradition of loyal service to both Crown and Country began in the 18th century and continues in the 21st, standing as a noteworthy exemplar of Canada's unique army. A key component of the Ranger tradition has always been the practice of promoting key leader education and command literacy and it is therefore appropriate that this volume, which aims to assist in that tradition, be dedicated to my Ranger family.

TABLE OF CONTENTS

Preface...i
Foreword..iii
Introduction..1

The True Nature of War..5
    Introduction...5
    Summary and Conclusions...16

The Utility of Theory..19
    Introduction...19
    Summary and Conclusions...29

Military Theory as Social History..33
    Introduction...33
    Society and Military Theory..35
    Conceptual Grafting..38
    Summary and Conclusions...50

War on Land..53
    Introduction...53
    Philosophers...56
    Theorists...59
    Strategists...72
    Summary and Conclusions...84

War at Sea..91
    Introduction...91
    Theorists and Strategists..94
    The World Wars...102
    The Submarine and Naval Aviation..104
    Summary and Conclusions...105

War in the Air..111
    Introduction...111
    Theorists and Strategists..114
    The World Wars...120
    Cold War and Nuclear Age..124
    The Air Power Decade...132

Summary and Conclusions..134

The Theory of Everything..139
Introduction..139
Revolutions in Military Affairs..141
Manoeuvre Warfare (MW) Theory..153
Misunderstanding Charles Darwin...158
Summary and Conclusions..161

The Future of War..167
Introduction..167
Electrons vs Blood..175
Clash of Civilizations..179
Future of Air Power..180
Conclusions..182

Nothing New Under the Sun..187
Introduction..187
Connections..188
Summary and Conclusions..191

Coda..201
Bibliography and Suggested Reading..207
General Index..223
Index of Personalities..225

PREFACE

Some time ago, I wrote a book, *Praxis Tacticum*, to assist junior leaders improve their tactical skills and understanding of their chosen profession. The Prussian general and theorist Carl von Clausewitz famously said that war had its own grammar but not its own logic. My earlier volume was aimed at Clausewitz's 'grammar'. This volume is in some ways the companion to Praxis. Although worthwhile for all military professionals and students of war, the purpose of this volume is to assist more senior leaders, students of military history and thought, and those who study war above the tactical level.

Humans are fascinated by war. The history of humanity is the history of war. It is a sad truth. Throughout humankind's history, we have regaled each other with stories of war and warfare. From the epic Greek myths and Norse sagas to modern day big screen movies, we cannot seem to get enough of war, at least as stories. And yet, we know that war is inherently a bad idea. It is evil. It is a form of collective madness. By its nature, war is destructive and cruel, unworthy of our better selves. To paraphrase Abraham Lincoln, war breaks the bonds of our affection. It does not always speak to our better angels. But having said that war's nature is both destructive and cruel, collectively, what do we really know about war's nature? It is this question which impels this book.

During my four decades spent as a Canadian armoured cavalry officer, I spent almost of third of it as an instructor. My own professional military education (PME) did not really begin until I was a young major sent to attend the two-year German Army Command and General Staff Course in Hamburg. Naturally, I did not appreciate how lacking my PME had been until I began my studies there, but it was not long before my Bundeswehr instructors ignited an inner passion to understand my chosen profession at levels, I was unaware existed. What follows in *Strategia* is the outgrowth of this passion and my own subsequent intellectual quest to better understand the military

STRATEGIA

arts and sciences.

    The title, from the Greek (στρατηγεία), encapsulates what this investigation entails. Students of military strategy and policy too often have no foundations upon which to build their understanding of those subjects, nor how they are so deeply entwined. This study offers such a foundation, thereby allowing a better understanding of military thought and theory. But reader beware! This is not an all-encompassing deep dive into the subject of strategy and military theory. The best way to describe it is that it is a primer, in the way that grade school children used to be given a primer to help them learn to read. Once introduced to the subject, the student, now armed with the basics, has the prerequisite tools to investigate further, to metaphorically drink more deeply of the mythical Pierian Spring.

    Finally, this work introduces many of the major works and theorists in military thought and theory as well as some of their most profound impacts on the conduct of war and the course of history. One cannot master military history without being familiar with these theories and theorists and their ideas. But what is presented here is not enough and I strongly encourage any serious student to seek out the original works of all the military philosophers, theorists and strategists covered in this volume.

Colonel (Ret'd) Charles S. Oliviero, CD, PhD

FOREWORD

I am honored to write the Foreword for Colonel Charles Oliviero's Strategia. "Chuck" and I are kindred spirits in at least one crucial respect: we both deplore the tendency among Western militaries to treat wars as battles writ large; America is not the only modern state to have a "way of battle" instead of a "way of war." Both Chuck and I have dedicated the bulk of our educational careers to exposing and redirecting that tendency. We have endeavored to orient the perspectives of contemporary military and policy practitioners to see armed conflict more holistically, as a function of not just military dynamics but also social and political ones. I am confident Strategia, though modestly described as a primer by its author, will move its readers that much closer toward our mutual goal.

Of the many authors who might have written this book, Chuck is absolutely the best qualified. He has taught military thought and theory for decades, drawing upon his ample experience as a military practitioner, to clear away some of the rhetorical "fog" that obscures history's most influential theories and theorists. After three decades of service as an armoured cavalryman in the Canadian Army, Chuck retired at the commendable rank of colonel. His professional military education includes the Canadian Army Command and Staff College, the Canadian Forces National Defense Studies Program, as well as the two-year-long German Armed Forces Command and General Staff Course (Generalstabslehrgang der Bundeswehr).

He is also highly accomplished linguistically, as he is fluent in French, German, and Italian, all of which he put to good use during his many years of teaching military thought from Machiavelli to Clausewitz at his alma mater, the Royal Military College of Canada. Chuck also served with distinction from 2006 to 2016 as an Adjunct Professor at Norwich University School of Graduate Studies for the Military History program. Rarely should we expect to find such an optimal combination of academic qualifications and military and teaching experience. Rarer still is it when that combination yields so effective a primer of armed conflict and military thinking.

Dr. Antulio J. Echevarria II
Professor of Strategy,
and Editor-in-Chief if the US Army War College Press

> "A little learning is a dang'rous thing; Drink deep, or taste not the Pierian spring ...

Alexander Pope
An Essay on Criticism

INTRODUCTION

I HAVE SPENT DECADES STUDYING AND TEACHING this subject, and over these years have noticed that some students did better if they had an interpretive guide, or primer, to help them wade into the 'deep end of the pool' of military theory. It was with this thought in mind that I wrote this book, to invite the reader, whether student, military professional or interested lay person, who is interested in military thought and theory to enter that pool. I have written this study as an advanced primer. It is not intended to replace any of the primary sources that a serious student must read. You cannot understand Clausewitz by reading someone else's condensation of *On War* (*Vom Kriege*). Like the mythical Pierian Spring, a condensed version of Niccolò Machiavelli's *Art of War* (*Arte della Guerra*) can be a dangerous thing. But having someone simplify some of the meatier ideas can certainly aid in achieving an understanding and if it whets the appetite, and leads to further study, then so much the better. Consider *Strategia* as your reader's companion or guide to military thought and theory.

    I have broken the guide into three parts: the first considers the various aspects of the nature of war. Is war a purely military activity? Does its nature change over time? Here we investigate how theory governs our understanding of war as it does all aspects of life. The issues discussed include definitions, the importance of language, the utility of theory in general and of war as both an art and a science. It offers an exposition of military thought and theory as a component of social history. The second part comprises military thought and theory regarding war on land, at sea and in the air respectively. The major philosophers, theorists, strategists, and practitioners are introduced and discussed as are some of their thoughts. The last considers whither military thought and theory are headed and consequently, what the future of war and warfare might be.

    In the second part, war on land, at sea and in the air, the relatively small number of individuals involved in naval and air power theories permits study in great granularity; however, the investigation refrains from delving into detail lest it detract from the intent of maintaining a holistic view. Such detail is simply not possible for land warfare due to the volume of material and so 'War on Land' is a broad survey. 'War at Sea' discusses Alfred Thayer Mahan, and Julian Corbett, primarily. Herbert Richmond and the French *Jeune École* are also briefly reviewed to appreciate not just the development of naval warfare but also the influences that naval theorists have had upon land warfare and vice versa. Similarly, the chapter 'War in the Air' discusses Giulio Douhet, and William Mitchell primarily with a few words on Hugh Trenchard and John Boyd and John Warden. There is a short discussion on the Cold War to help frame the development of aerial warfare under the aegis of the 'nuclear umbrella' as well as to investigate the influences and impacts those aerial theorists have had with surface warfare and vice versa.

My proposition is straightforward: More than two millennia of investigation has brought Western society only marginally closer to understanding the nature of war. Broadly speaking this premise is revealed in three ways: First, in common parlance, the concept of war has become nearly meaningless. Second, military professionals are not conversant in more than one or two military theories, nor can they draw a distinction among military philosophers, military theorists, and military strategists. Third, there is little or no appreciation in Western society that at best modern military theories are artificial hybrids of many older theories or worse, they are syncretic amalgams of multiple disjointed ideas. At their worst, some modern theories are pure fantasy wrapped in authoritative language. Although these latter two concerns, professional ignorance and nonsense pretending to be theory, are distinct issues they are so intimately related and affect each other such that they are best discussed together as a single manifestation of the book's premise.

As an old cavalryman, allow me a word of advice to the serving military leaders who may be wondering why such theoretical study is even necessary. A well-worn military maxim reminds us of the need to leaven experience with study. The great Marshal Maurice Comte de Saxe once famously remarked that a mule that had served twenty campaigns under Caesar, would still be but a mule. Therefore, experience does not by itself make a great commander. Whatever innate skills and abilities he or she may have must have been improved not only by practice, but also by study. Thus, without study, a breach may exist between personal experience and broader history. This guide looks to help close that breach.

In closing, at times you may wonder if I split hairs. Let me explain in advance why I believe that nuance is critical to gaining new insights into war's nature. The student of war must be an educated investigator. The subject matter is too important not to be and ignoring subtle differences may have important consequences. An analogy with the study of the law is appropriate. Lawyers, judges, and legal scholars all learn early that in the study of jurisprudence, intent often is a determining factor in

INTRODUCTION

finding guilt or innocence; intent is key, nuance governs justice. When someone kills a fellow citizen, whether there was intent, how early the intent and how that intent was manifested all come into play. The victim is dead irrespective, but how and why the death occurred remains critically important to those charged by society to find justice. As far as understanding the nature of war, consider Winston Churchill's insightful comment that "Battles are won by slaughter and maneuver. The greater the general, the more he contributes in maneuver, the less he demands in slaughter." In that vein, the more we understand of war, the less we will demand in slaughter.

"Nothing so comforts the military mind as the maxim of a great but dead general.

―

Barbara Tuchman
The Guns of August

THE TRUE NATURE OF WAR

Introduction

WHAT IS THE TRUE NATURE OF WAR? The question is deceptively simple; but the answer is not. In fact, whether the true nature of war can be understood at all is moot. However, there is value in the pursuit. The premise here is that despite more than two thousand years of active consideration, Western society has gained little in its understanding of the nature of war. Such an assertion is not entirely new. Others have made similar claims.

The Chinese philosopher Sun Tzu wrote his treatise *Art of War* circa 500 BC. Since then, humankind has accumulated a great deal of experience regarding war; but this accumulation has not necessarily been helpful in gaining insights into its nature. Unquestionably, war has become more complex over time. Sticks supplanted bare fists just as slings and arrows have given way to computers and directed energy weapons. Muscle power has given way to industrial and chemical power, which has given way to electrical power. The conduct of war has become multi-dimensional. Humanity has learned to slip the bonds of gravity just as it has learned to live and fight under water. Certainly, war is ever more complex. But this complexity has never referred to war's *nature*, only to its *conduct*, the former being immutable while the latter is constantly changing.

Our understanding of the nature of war is akin to the proverbial investigation of an elephant by six blind men. Each man, after investigating a different part of the same animal, variously describes it as a tree-trunk, a snake, a rope, a wall, a spear, or a fan. In *The Six Blind Men of Indostan*, the American poet John Godfrey Saxe (1816-1887) left the reader with this concluding moral:

So oft in theologic wars,
The disputants, I ween,

STRATEGIA

> *Rail on in utter ignorance*
> *Of what each other mean,*
> *And prate about an Elephant*
> *Not one of them has seen.*

Thus, is war really an extension of politics by other means as Clausewitz asserted? If so, then the nature of war must be political. Alternatively, perhaps war is really an expression of greed based on economic and class inequities as Karl Marx declared. If so, then the nature of war must be economic. Perhaps war is a social disease, as some social scientists have come to believe. If so, then the nature of war must be social. Perhaps humanity has a genetic need for war. The fact that a couple of millennia of recorded study have not resulted in a definitive answer to the question of war's nature would lead a reasonable investigator to conclude that war must surely contain at least some aspects of all the above premises. War is surely political, social, economic, and more. The obviously multifaceted nature of war circumscribes all these qualities and likely many more. War is among the most elaborate and complex of all of humankind's pluralistic activities, from simple tool usage, to language, to the building of new societies. It is also the most destructively absurd. Could this absurdity be because humanity does not understand it?

It has been my experience that most military professionals are not appropriately conversant in Military Theory, nor can they draw a distinction among military philosophers, military theorists, and military strategists. This deficiency in their professional educations stems from ignorance. Surveys have demonstrated that most military professionals cannot name more than a few military thinkers. This circumstance has led to the misconception that there have been only a few such people throughout history. Most are familiar with Carl von Clausewitz – but only one in fifty have read *On War*. Most have heard of Henri de Jomini, but only one in a hundred have read *Art of War*. Some have read Sun Tzu. But in-depth knowledge of other theorists is so sparse as to be statistically insignificant.

The curriculum of the world's oldest staff college, the German *Führungsakademie de Bundeswehr* only teaches Clausewitz. Although Sun Tzu, Colmar von der Goltz and Jomini are mentioned, they are not taught. Canadian Forces College in Toronto does not teach Military Theory at all, *per se*. The British army command and staff college does little more than provide a grocery list of major theorists. The Australian command and staff college does likewise. An investigation of the curricula of national war colleges offers different insights: The US Army War College, the British College of Defence Studies, and the US Air War College all offer studies in military theory. However, the number of officers selected for study at these high-level colleges is less than 1%. The civilian equivalency would be comparing the number of graduates from high school with the number of students who finish their doctorates. For military professionals such ignorance manifests itself as a lack of knowledge of military thought. With only

THE TRUE NATURE OF WAR

Sun Tzu, Clausewitz and Jomini to draw upon it is perhaps understandable that many military professionals do not appreciate the full breadth and depth of the subject. As a direct consequence of this poor historical appreciation, the subtle distinctions among philosophies, theories and strategies are left unnoticed.

Western Military Theory is an admixture of differing national beliefs and assumptions, often forming contradictory ideas. This heterogeneous mixture is not in itself the problem. Just as some addicts know that their habit is injurious to their health but believe that they will not fall victim to its effects, the same sort of ignorance or denial is present here. The people who are mixing and matching bits of discordant philosophies, theories, and strategies – even professional military officers – do not appreciate that their understanding of war is an artificial hybrid. In fact, military officers are among the worst offenders. As few officers have either the time or the inclination to study the mass of philosophy or theory underlying their national strategies, they conveniently accept the amalgams as presented in various doctrine manuals. The fact that this literature is not founded upon any single philosophy, is not a rationalized theory, nor is possibly even a workable strategy remains veiled and therefore misunderstood.

My friend Professor Allan English, who for years taught history and operational art to senior Canadian officers at the Canadian Forces College, shared his own experience on the subject. As a retired officer himself, he noted the reluctance of senior officers to study theory, many of whom held that abstract theories were not useful in the real world of war. The majority failed to appreciate that much of the operational art is based on abstract concepts like synchronization, integration, manoeuvre, or centres of gravity. Although Professor English's example is Canadian, the same attitude is found across Western militaries.

Reviewing results from a hypothetical survey of North Atlantic Treaty Organization (NATO) officers asked to describe the nature of war would uncover that the belief in the duality of war stems from Clausewitz, the current "golden-haired boy" of senior military leaders. But the NATO belief in an underlying set of fundamental principles of war stems from Henri de Jomini not Clausewitz. Importantly, these two theorists were at odds and their thoughts were in many respects *irreconcilable*. Clausewitz believed that there were no hard and fast rules in war. He insisted on the 'particularism' of war and combat. That is, he felt that each specific war was unique. He refused to accept the concept that there could be a set of eternal underlying principles, which governed warfare in the same way that there were principles in physics or chemistry. Jomini believed the opposite. His studies of Napoleonic battles convinced him of the existence of a set of principles that governed the conduct of battle almost as rigidly as the laws of thermodynamics govern combustion. Moreover, as we see in the table below, each country's military culture has interpreted Jomini to serve its own ends and has arrived at a different number of principles.

Note the different principles for something that is supposedly universal: three

7

THE PRINCIPLES OF WAR	
NATO Human Factors Selection and Maintenance of the Aim Freedom of Action Concentration of Effort Economy of Force Mobility Surprise Intelligence Simplicity Maintenance of Forces	**US Army** Objective Offensive Mass Economy of Force Manouevre Unity of Command Security Surprise Simplicity
UK/Canada Selection and Maintenance of the Aim Offensive Action Concentration of Force Economy of Effort Flexibility Cooperation Security Surprise Maintenance of Morale Administration	**France** Concentration of Effort Surprise Liberty of Action

for the French, nine for the Americans, ten for the British and the Canadians, and ten for NATO. The Germans, as dyed in the wool Clausewitzians who believe in the particularity of each war, would not agree that there were any at all. (Unless they were speaking as NATO officers, in which case they would agree upon ten!).

Consequently, the difficulty lies not only in the fact that Western Military Theory is a diverse amalgam of ideas. There is nothing wrong with selectively choosing bits and pieces of different, even opposing philosophies, theories, and strategies if it is done consciously. Properly done, this syncretic process becomes the familiar thesis, antithesis, and synthesis of Hegelian thinking. It is when the process is accidental, when there is no disciplined intellectual evaluation that the process becomes invalid. In the case where the accidental process results in a military theory or strategy, the product can be dangerous or even catastrophic.

In his commencement address at Yale University in June of 1962 President John F. Kennedy told the graduates:

> The great enemy of truth is very often not the lie – deliberate, contrived, and dishonest – but the myth – persistent, persuasive, and unrealistic.

THE TRUE NATURE OF WAR

Too often we hold fast to the clichés of our forebears. We subject all facts to a prefabricated set of interpretations. We enjoy the comfort of opinions without the discomfort of thought.

This intellectual sloth has been common in the realm of Military Theory; most Western military officers, sometimes quite senior, have lived their entire careers in ignorance of the intellectual underpinnings of their profession. Habitually and often disdainfully, they have consigned the study of their vocation to academics, thereby improperly relegating themselves to their profession's intellectual sidelines. I would go so far as to say that there has been a longstanding tension between soldiers and academics. My personal experience has been that most officers think of theory the way Hermann Göring thought of culture. Göring reputedly said that when he heard the word culture, he released the safety on his Browning pistol. Such behaviour is foolhardy at best and dangerous at worst. Unlike most professions, the pursuit of war is not the exclusive domain of any single group; war is not proprietary. Although militaries like to believe that professionals control war, amateurs abound. Historical examples where amateurs beat professionals are plentiful. The Bolsheviks defeated the Imperial Russian Army; Mao Zedong's peasants overthrew the American-backed Chinese Nationalist Army of Generalissimo Chiang Kai-shek. It behooves all involved in military affairs –particularly professional military officers – to appreciate that understanding the nature of war is an elemental key to controlling it. Those with a better understanding of war can and often do defeat those who are better-trained, better-prepared, but less-informed.

Sometimes it is not intellectual sloth but circumstance. In multinational settings, the mixing of terms and concepts can be subtle, and insidious. Alliances bring together armed forces of different nations. NATO is a perfect example where more than two dozen nations with almost as many languages have been working for over half a century to come to common understandings of war and warfare. These nations may use terms and concepts differently. The aim of any alliance is to allow disparate nations' forces to fight alongside, or as part of, other nations' forces. The military calls this 'interoperability' and it can run the gamut from common ammunition to agreed-upon languages of command. Regular military exercises strive to improve this ability. Allies exchange officers and allow each other's officers and soldiers to attend each other's courses and staff colleges. Thus, in time of war, allies can fight alongside each other using common and agreed to methodologies and procedures.

This is how interoperability works in principle. However, owning a dictionary does not assure literacy and simply establishing common procedures does not guarantee common understanding. Not all countries share political or military philosophies. Allies may have distinctly different theories of war. They may have overlapping philosophies and theories, but this overlap may not extend to strategies. Were we to survey senior military officers to describe the nature of war, it would be a sure bet that

STRATEGIA

the majority would reply that war's nature contains a dichotomy. Whether they spoke in terms of *ordinary* and *extraordinary* forces or referred to the *moral* and the *physical* would depend upon whether they were more inclined towards Sun Tzu (ordinary/extraordinary) or Clausewitz (moral/physical); but it would be a near certainty that in their training they would have been exposed to both.

One of the most fundamental issues we face is one of definition. What is war? What does it mean to be at war? These rhetorical questions can be deceptive. Western society uses widely the concept and vocabulary of war. Common usage would imply that the term war, an act that usually pertains to soldiers, fighting and killing, is clearly understood. War is, after all, a broadly shared human experience. It is a fundamental human endeavour, a social behavior; it is adversarial, highly dynamic, and complex. All human cultures have known war. All societies have had many generations of practice with war. This enormous repository of experience has led to a widespread acceptance that humanity understands war; but this is not so. If Western society overuses, misuses, or abuses one concept, undoubtedly that concept is war.

Canada is not currently at war. Yet, it is practically impossible to make it through an entire day without seeing, hearing, or using words, ideas, or concepts about war. Western society appears convinced that it is, in some way or other, continually at war. Consumers are inundated daily with the analogies, metaphors, and images of war. North America is in the middle of a war on poverty. The Western world has been conducting a war against drugs for over three decades. There is an ongoing war against illiteracy. The American president has declared a war against terrorism.

As further proof of the widespread misuse of bellicose words and concepts, consider that many professional groups have adopted military vocabularies. Stockbrokers fight desperate actions in the financial trenches to secure the perimeter of corporate profits. Medical doctors launch counterattacks against AIDS and cancer. Dentists bombard their patients. Barristers gird their loins for combat in the arena of jurisprudence.

Understandably, most of this misuse is hyperbole. It can be exciting for people to use the aphorisms and metaphors of heroic struggles. To most of the populace, this language somehow makes the mundane seem important. In a world of bellicose adages, the drudgery of making money takes on the importance of a Homeric quest. Selling the latest computer software becomes as urgent as winning a pitched battle. Such language lends an air of authority to ordinary tasks, irrespective of their actual importance. During the 'cola wars' Coke and Pepsi never actually fought each other. Most of this misuse is merely literary license. There is also a sense of clarity when at war; the enemy is defined, and the nation united in a struggle against a common foe. But calling something a war does not make it so.

Some of the blame for such extensive misuse is certainly due to the inherent flexibility and evolution of the English language:

For example, the root of the English word 'war', *werra*, is Frankish-German, meaning confusion, discord, or strife, and the verb *werran* meaning to confuse or perplex. War certainly generates confusion, as [Carl von] Clausewitz noted calling it the "fog of war", but that does not discredit the notion that war is organized to begin with. The Latin root of *bellum* gives us the word belligerent, and duel, an archaic form of *bellum*; the Greek root of war is *polemos*, which gives us polemical, implying an aggressive controversy. The Frankish-Germanic definition hints at a vague enterprise, a confusion or strife, which could equally apply to many social problems besetting a group; arguably it is of a lower order sociological concept than the Greek, which draws the mind's attention to suggestions of violence and conflict, or the Latin, which captures the possibility of two sides doing the fighting.

The present employment of 'war' may imply the clash and confusion embedded in early definitions and roots, but it may also, as we have noted, unwittingly incorporate conceptions derived from particular political schools.[1]

Languages are living things. They must live and grow, but not to the extent that words lose their ability to convey meaning.

These distinctions are not just semantics; they are important. If terms and meanings are confused, then there is little prospect of gaining any better understanding of what is being discussed. Rational debate is predicated upon the clarity and explanatory power of language. Understanding is a function of the capacity of carefully chosen conceptual terminology to clarify and to convince. As with the problem of Western society overusing and abusing the concepts and lexicon of war, so long as there is a lack of a complete understanding of the nature of war, there can be no hope of controlling or eliminating it. Unfortunately for the student looking for a better understanding of war and warfare, the theoretical literature is mostly in the realm of linguistics, psychology, and philosophy.

Consequently, we are left to investigate how and why we have these longstanding gaps in our collective understanding of war's true nature. To accomplish this examination of military thought and theory, we need some tools. We begin with a new model [Military Theory 'A'], which addresses the relationship among philosophy, theory, strategy, policy and tactics, techniques, and procedures (TTPs). The model offered below is a modest proposal: philosophy is the foundation of theory; theory is the foundation of policy; policy is the foundation of strategy; strategy is the basis of tactics and so on. Like building stones that rest upon each other, they form a pyramid and rest upon the ideas below them. Together they comprise Military Theory.

It is important to appreciate the taxonomy in the model above. Philosophy is not just a component of Military Theory; it is the bedrock of all theories. Likewise, there are manifold theories pertaining to war, all of which are critical components of Military Theory and all of which are founded upon one or more philosophies.

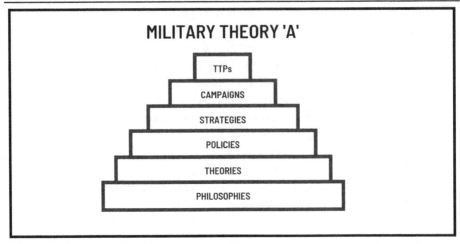

Subsequently, national policies are created by governments consciously, or otherwise, predicated upon these philosophies and theories. Strategies are thereafter necessarily the products of such policies and although they are crucial to the conduct of war, they remain products built upon the base of manifold theories. Thus, the components of Military Theory are not co-equal. Order and progression are critical. In the hierarchy of Military Theory philosophy, being the foundation, is more important than theory. In turn, theory is more important than policy, strategy more important than campaigns and so on.

The model above at 'A' offers a fresh concept of how philosophies, theories, strategies, policies and so on interact. Within the context of war, it is important to distinguish among these terms. Philosophies are the underlying sets of beliefs upon which a society, including its armed forces, bases all that it does. Theories are compilations of premises to aid in understanding. Policies define the parameters within which leaders can behave. Strategies are plans of action to achieve goals. Campaign plans establish methodologies and courses of action to enact the strategies and Tactics, Techniques and Procedures (TTPs) are the low-level means of achieving limited objectives.

Next, we need consider a new paradigm, or model, a way of better appreciating the interconnectedness between and amongst all aspects of military thought and theory. I call it "The Matrix of Military Theory". This complex matrix embodies the interconnectivity of philosophies, theories, policies, and strategies as well as tactics, doctrine, war, and warfare. Together, all these components combine to create this matrix. But like all models, the matrix is useful only to a point. It helps explain that military thought has not been linear. It is valuable to express the idea that like biological RNA, all philosophies, theories, policies, strategies, and tactics as well as myriads of other components of Military Theory combine and recombine, offering infinite alternatives rather than a single 'true path'. Naturally, it has limitations. The concept of a matrix does not do justice, perhaps, to the multi-dimensionality of the connections. The linkages of time and space, for instance, may suffer somewhat since the model of a matrix implies at least a minor linear progression where one may not exist. Nonetheless, accepting the shortcomings, the concept of a matrix remains a highly useful tool to act as the paradigm within which to pursue our investigation.

Next, consider the accepted levels of war. For more than a millennium, there were but two accepted levels of war: the tactical and the strategic. Beginning in the late 19th century, a third level crept into military thinking and there are now three universally accepted levels: from highest to lowest, they are the strategic, the operational, and the tactical. Originally based on the Napoleonic Wars, it was the Prussians who formulated this tripartite linkage although there is widespread proof that the Soviets developed it to where we understand it today. Consider World War II. The Allied High Command required a strategy that encompassed all theatres of war from the Arctic Circle to the South Pacific. Then, each of those theatres needed commanders who could translate the Allied strategy into operational victories using the dozens of army corps from all the contributing nations. Further down the chain, at the tactical level, commanders from army corps down to infantry sections had to fight engagements, battles, and skirmishes.

Clearly, these levels are neither independent nor autonomous. In other words, they are intrinsically connected to each other, and they tend to blend from one to another without sharp lines of demarcation. They also are defined differently in different military cultures. Although all agree that the tactical level begins with individual soldiers and the strategic is the highest level, countries differ on where the operational level starts and ends. What everyone does agree on is that the operational level is the connective level; it is the indistinct area between tactics and strategy that connects them in both directions. For our purposes, the simplest way to think of operational-level actions is to consider that they are best described by tactical actions, which have strategic outcomes. But I would like to expand this understanding by one more step. Contemplate a fourth level. If we consider that above the strategic level, there is an ephemeral, non-physical level, then let us call it the theoretical level. For those who have read Plato's *Republic* you will be familiar with his Allegory of the Cave. I have no desire to engage in a philosophical dissertation of Platonism, but his allegory is useful. In the allegory, a situation is described where people live in a cave all their lives and who only experience reality based on the shadows cast upon the wall of that cave. The shadows are obviously not accurate representations of reality but for those in the cave the shadows are all they must interpret the real world.

In the allegory, the philosopher is the cave dweller who is freed from the cave and comes to see that the shadows are not reality at all, that they are just representations of that reality. Looking at this allegory is how we should understand this fourth, theoretical, level of war. To the student of war, this higher level represents a deeper understanding of war's nature, rather than simply seeing only the physical manifestations present in the real world. In other words, using these and other models we can seek a deeper understanding of what the physical manifestations can tell us about the true nature of war.

With some of the foundational stones laid, and before moving on to appreciating the utility of theory, we need to take a short detour. Above in model 'A', I offered you a taxonomic model of Military Theory. It was the basis of what I called a complex matrix. Look again at that model. I have gone to some effort to describe the model as a *taxonomy*, a system of describing the way in which things are related by putting them into groups. I was being somewhat disingenuous but did so on purpose. Although the pyramidal model in 'A' is indeed a taxonomy, there may well be a better way – albeit more nuanced way – to describe the next step, the creation of a complex matrix. That better way would be to consider the component parts of the pyramid as an *ontology*, or a collection of related taxonomies.

Now, if we consider the model in 'A' as an ontology rather than using a simple hierarchical pyramid, a better representation of the model might be as we see in [Military Theory 'B'] at right, a simplified matrix. I am not saying that the second model is a better representation than the first, only that it *might* be. The second version is illustrative only. So, what is the point? Well, the point is that to understand either model you need to appreciate that the models are about relationships. Understanding my models (and the nature of war) is all about understanding the components that make up war and the vast array of relationships between, among and within these components. And that is why I suggested that perhaps a better way to look at the model *might* be to look at it as an ontology rather than as a taxonomy.

You may be wondering what the difference is between a taxonomy and an ontology. To begin, remember that a taxonomy classifies hierarchical relationships. In model 'A', we see that philosophy is a necessary foundation to any part of the model above it (theory, policy, strategy *et al*). But you might legitimately argue that the taxonomy that was presented only offered to categorize items within one facet of war and that war is multi-faceted. No argument there. I used the model in a simplified form to make it easier to understand at an introductory level. But what if we wanted to understand the model in higher order domains? Then, we would need to consider the taxonomy

THE TRUE NATURE OF WAR

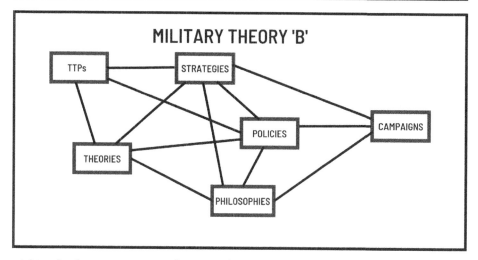

within the larger structure of an ontology, which can provide more information regarding the relationships in the model.

But what is an ontology? The simplest way to look at an ontology is to consider it as a collection of related taxonomies, in the way that a clan is a collection of families. For instance, we might want to consider the relationships that could be developed if we had multiple versions of the pyramid, with each one representing a different country or military. The relationship between philosophy and theory in France might be subtly different than it might be in Germany. And then there would also be the relationships between and among the various building blocks across the various nations. It would not take long to create a highly complex series of connections – the so-called "Complex Matrix of Military Theory." Demonstrating even a few of those connections makes that clearer.

Since ontologies depend upon taxonomies and are necessarily difficult to develop – because they compel the designer to have a holistic understanding of the subject – they quickly grow too complex to display easily. Add to this complexity a need to understand the vagaries of language, culture, political structures and the other human variables and the matrix can quickly become too complex to be of any use to anyone except old cavalry colonels like me who enjoy woolgathering.

Below, as Model 'C,' there is a simplified attempt to display an ontology of models of military theories. Such a model might not help you at this stage, but for those adventurous enough to move beyond the simpler models, using what Albert Einstein called 'thought experiments', the process of moving from A to B to C may reward you with new insights.

Last, the breadth and depth of this investigation is highly ambitious. Historical surveys can be useful tools. However, covering a historical period that spans over two thousand years forces difficult choices. There has been no attempt to produce

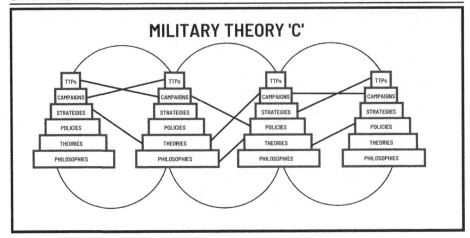

a classic battles-and-leaders narrative. Instead, this work forms an analysis of how societies thought about war. It is a survey of historical ideas. Obviously, we cannot discuss all of those who have had an impact upon Western society's understanding of the nature of war. The examples chosen are those that have been unique in their impact, most representative of a certain trend or school of thought or most influential to the profession of arms. The survey proceeds thematically rather than purely chronologically. However, within each theme the narrative does pursue a chronology.

Summary and Conclusions

Several tendencies drive the need for this investigation: a society largely uninformed by historical knowledge; the academic deficiencies of military officer corps that are products of generations divorced from their intellectual roots; and the tendency in military cultures to use sweeping generalizations and simplistic explanations for war without appreciating the danger of so-called 'very big ideas' that have more in common with the facile marketing industry than they do with the art of war. These tendencies lead to superficial understandings of the nature of war. Western liberal democracies, with their innate belief in positive futures, are highly disposed to believe that some unique future strategy, based on who knows what, will save them from an upcoming disaster brought on by unpreparedness: We need not concern ourselves with climate change, *something* will save us. We may be running out of fresh water, but *someone* will figure out how to solve this problem. Similarly, newly presented philosophies, theories or strategies are presented much like new cleaning products or the latest model of automobile and however elegant or attractive they may be, historically illiterate cultures cannot hope to understand these concepts if, in JFK's words, 'We enjoy the comfort of opinions without the discomfort of thought.' For example, the siren song of technology is heard everywhere in military circles. Yet, the same leaders who claim that better technology has and will continue to alter the nature of war, glibly profess to base their beliefs in, and continue to draw quotes from, the writings of men like Clausewitz, Jomini and Sun Tzu, all of whom were quite clear that war has an immutable nature and as a profoundly human escapade, is governed more strongly by emotion than by science.

In our first step towards investigating the unknown nature of war, we have built a series of models to assist us. But let us be clear. None of the models presented lay claims to be a new theory of war. Rather, they are tools to aid in our understanding of Plato's shadows on the cave wall, of coming to grips with the fact that the nature of war seems immutable while at the same time the character of war changes continuously. Can we draw any conclusions so early in our investigation? Yes. First, the study of Military Theory suffers from a lack of institutional discipline. Most Western militaries claim to be professions and yet there are few if any universal standards, no worldwide agreements on what is true and what is not, no international bodies charged with adjudication of the profession as it relates to its own education. To help us clear some of the muddle, we now also have a framework, the paradigm of a complex matrix. All the philosophies, theories, policies, strategies *et al* interconnect to make up his matrix. The manifold aspects of Military Theory must be understood both individually as well as collectively and until such time as Western military culture understands this need, the conclusion of the investigation would seem incontestable: Despite more than two thousand years of active consideration, we remain ignorant of war's true nature.

Notes

1. Alexander Moseley, "The Philosophy of War" available from https://iep.utm.edu/war/.

> "At the root of Alexander's victories, one will always find Aristotle.

Charles De Gaulle
The Army of the Future

THE UTILITY OF THEORY

Introduction

GENERAL DE GAULLE WAS RIGHT. Aristotle laid the philosophical foundation by teaching his pupil about the nature of the universe; Alexander then created his own military theory, which he then translated into strategy; he did not simply leap upon Bucephalus and set out to conquer the world.

We now turn our investigation to the utility of theory in general and then consider the various relationships within our Complex Matrix. The investigation highlights a particular need: Society in general, and especially military officers, need a better understanding of Military Theory. Only by gaining this better understanding can the profession of arms fulfill its obligations, not only to society, but also to itself. Military Theory is an exceedingly broad topic. Hence, it is best to begin with definitions. Returning to the foundational model 'A', we can see that the domain encompassed in it is unworkably large, particularly if our study focuses primarily upon war, for arguably the military community has lost ownership of this subject. Even cursory research demonstrates that sociologists, anthropologists, and psychologists have hijacked modern Military Theory. The domain must be narrowed. If we establish Military Theory as the universal domain, then the subdivision of all the various related areas of study may be broken out as sub domains. The potential to divide and subdivide this domain is almost unlimited. Therefore, classification becomes important.

A negative example of division and classification, one that adds little to the understanding of war, can be found in the work of Gérard Chaliand, a Belgian-French expert in geopolitics who has published widely on military strategy, who has argued: "It is important to develop, however difficult and complex it may be to do so, a typology of wars, beginning with the observation that there have been manifold concepts of warfare and its nature throughout history."[1]

Chaliand offers one such typology:

Ritualized Wars
Wars With Limited Objectives
Conventional Wars of Conquest
Mass Wars
Wars Without Quarter

The British philosopher Alexander Moseley, in his 2002 *A Philosophy of War* provides another list:

Animal Warfare
Primitive Warfare
Civilized or Political War
Modern Warfare
Nuclear Warfare
Postmodern Warfare

Unfortunately, after a great deal of time and effort labelling and ordering wars in this way, at least from a military perspective little is gained in the understanding of the nature of war. So, although it may well be an interesting intellectual pastime, like arranging stamps in an album, the creation of a typology should not be mistaken for the achievement of an understanding. Classifying different types of internal combustion engines is not the same as understanding how and why they work.

Originally, Military Theory was almost exclusively war centred. It focused on the fighting of wars between nations and how such wars could be won. The historical literature bears this out. From Sun Tzu (*Art of War*) to Niccolò Machiavelli (*Arte della Guerra*), to Carl von Clausewitz (*On War*), and Antoine-Henri Jomini (*Précis de l'art de la guerre*), the most widely read and studied military writers all referred primarily to war between states. This limited view has been what most soldiers, past and present, have mistakenly understood to be the whole domain of Military Theory. It is wrong. As the model at right [Military Theory 'D'] demonstrates, the domain is vast, and the war-centred aspect is but one of many sub-sets. War-centred Military Theory, the dominant original, was eventually broadened as other theories joined the domain. Over time the study encompassed wider and wider ranges of military activity. Further, the study did not confine itself only to war; it was also concerned with other uses of power and military force.

This broadening of the domain led to the addition of subsets only remotely concerned with war: the avoidance of war became part of the domain as did peace research as did the study of eliminating all forms of violence from society. In the end, there was no constraint upon what belonged to Military Theory *per se*. As a case in

THE UTILITY OF THEORY

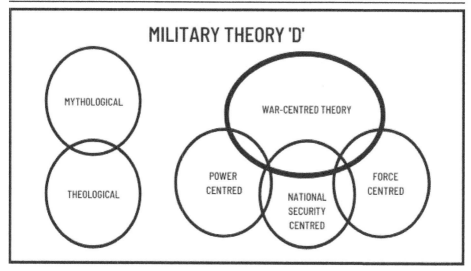

point, the combined reference libraries at Kingston, Ontario of the Canadian Army Command and Staff College and the Royal Military College form perhaps the best military collection in Canada. A broad search of their combined collections turns up, of course, the writings of Sun Tzu, Jomini, Clausewitz, *et al*. What is interesting, however, is that the search of their collections on the topic of 'violence' also turns up almost 100 titles, all of which were interesting studies, but all of which were *sociological* texts aimed at achieving world peace or social harmony. A similar search for 'military theory' called up Jomini and Clausewitz almost exclusively since authors like Sun Tzu and Machiavelli were found under philosophy and political science respectively.

We must restrict our investigation here almost exclusively to war-centred theory. All other subsets, such as those that concern social scientists, political scientists, psychologists, economists, and theologians are not investigated. With this constraint in mind, the models portrayed above as [Military Theory 'A'] and [Military Theory 'D'] combine to become [Military Theory 'E'] below: Hence, notwithstanding its manifold variations, when the term Military Theory is used, the discourse will concern itself only with war-centred philosophies, theories, strategies and so on.

In 'The Unknown Nature of War' above, three manifestations of the premise that little progress has been made in the understanding war were presented. The first manifestation was that the concept of war was poorly understood. Likewise, there are issues with what constitutes a theory. What is theory? Like the word *war*, the word *theory* is misunderstood and misused. Each year, professors admonish freshman engineers not to use *theory* when *theorem* is what they mean. Professor A.J. Barrett used to remind my freshman engineering class: 'Gentlemen, Dr Einstein had a theory. Monsieur Fermat had a theorem.' But the professor was fighting a losing battle. Increasingly in popular culture, the former is used when the latter is meant. Agatha Christie had her punctilious Belgian detective misuse the word constantly. Sir Arthur

STRATEGIA

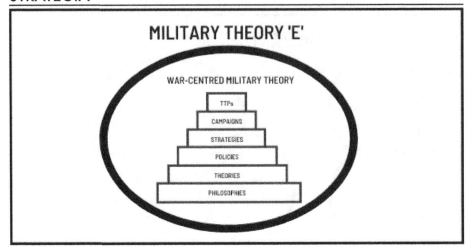

Conan Doyle did likewise with Sherlock Holmes.

Most people use the words *theorem* and *theory* interchangeably. They are not the same; the former is a proposition whereas the latter is a framework for a belief system. A theory is a supposition or system of ideas explaining something, especially one based on general principles, like the exposition of the principles of a science or the collection of propositions to illustrate the principles of a subject. Most live their lives in blissful ignorance of any theories, but consciously or not, everyone uses theories every day. They provide intellectual structure. They are invaluable aids to clear thinking and assist in developing rules. Understanding the world around us would not be possible without theories. A rudimentary understanding of physics, for instance, is fundamental in the building of a bridge. Some knowledge of the nature of wood is a prerequisite for carpentry.

Clausewitz felt that the purpose of theory was to educate. Only through education can you hope to understand war. Only through understanding can you hope to master this most violent activity. Professor Williamson Murray shares this opinion; "theory, with a small t, does have its place as an organizing principle – to catalogue our thoughts and to extend our understanding of the complex, ambiguous, interactive phenomena that make up the real world. Particularly in the world of military institutions, theory about war (or theories about war) provides some direction to thinking ..."[2] Thus, in all facets of human life – including war – mankind's thoughts have been ordered by theories.

As far as understanding the nature of war, at least since Sun Tzu there has been no lack of explanations. Many philosophers, theorists and strategists have claimed to uncover war's true nature. Jomini bragged: "The science of strategy consists, firstly, in knowing how to choose a theatre of war well ... The art, then, consists, of the proper employment of one's troops upon the field of operations ... The whole science of great military combinations rests upon these two fundamental truths"[3] Unfortunately, such

THE UTILITY OF THEORY

positive assertions have not guaranteed understanding of war's nature. As with Plato's Allegory of the Cave, such claims are shadows that reveal certain aspects of war's nature rather than a direct and complete observation of the phenomenon itself.

Although not uncovering the secret nature of war, each philosopher, theorist, or strategist did bring meaningful insights to the discussion. Some did so better than others. A handful of men like Machiavelli, Clausewitz, Jomini, Alfred Thayer Mahan, Julian Corbett and Giulio Douhet became both famous and influential. The remainder made their impacts by training subordinates, like August von Gneisenau's influence upon Clausewitz or Dennis Hart Mahan's upon his more famous son. Conversely, like Philip Howard Colomb, they sank into obscurity. Whatever their influence, in Platonic terms their explanations remain mere shadow cast upon the wall; none have answered definitively the question of what war is or have revealed completely its true nature.

If we have made so little progress, then why formulate theories of war at all? Why bother? Theories are often seen negatively. They have often taken on negative connotations: Something may work in theory but not in real life. The sentiment is not uncommon among practically minded people and especially amongst military professionals. Particularly amongst combat officers, theory is given little credit or credence and consequently, little study. Disregard for theory is commonplace and it is arresting to think that this ignorance holds true for the great majority of the world's soldiers. To them, military actions occur because that is the way it has always been. Soldiers march in step because it is 'well known' that they have always done so. Ironically, this is not true for they have not always done so. It may seem obvious that armies meet each other in battle. Soldiers fight other soldiers. Sailors sink each other's ships. Pilots attack other pilots as well as their supporting armies and navies. To these men and women, so highly focused upon actions and results, worrying about obscure theories can seem irrelevant.

Such ignorance may be widespread, but it is by no means universal. There is ample evidence that those military leaders who read widely and studied their profession were frequently more successful than those who did not. Frederick the Great made the study of the military arts and sciences his life's work, the mastery of which was aptly demonstrated in his *The Instructions of Frederick the Great for His Generals* (1747). Likewise, the two officers in the US army with the largest personal libraries were George S. Patton Jr. and Douglas MacArthur, both of whom were outstanding battlefield commanders. The German General Staff was a collection of such well-read men. Officers like David von Scharnhorst, Helmuth von Moltke, and Hans von Seeckt, by virtue of their upbringing and military education were all well-grounded in military classics and military history. But they were in the minority. For generations, armed forces have maintained an anti-intellectual bias, preferring to focus upon teaching *skills* instead of *ideas*. The Canadian Forces Command and Staff College teaches only Clausewitz in any depth. This is true also of the *Führungsakademie*. The US Army

STRATEGIA

Command and General Staff College at Fort Leavenworth teaches both Jomini and Clausewitz but the study of the writings of Sun Tzu, Ardent du Picq, Giulio Douhet, Alfred Mahan *et al* is left mostly to their School of Advanced Military Studies and the upper-level War Colleges. Consequently, these teachings are reserved to the select few who will likely go on to achieve flag rank. Thus, most military officers remain ignorant of any but a rudimentary formal contact with military theories. The French, Italian, German, British, Turkish, Canadian, American, and Dutch armed forces, comprising easily the best and most widely educated officers in the world, are all similar in this respect; an unflattering judgement of the professional military education (PME) in all these military forces.

Clearly, theories are not redundant. They are essential to understanding. Based on fundamental philosophies, theories form the paradigm within which or upon which strategies and plans of action are formulated. But this study need not proceed in a formal linear fashion. The paradigm of a matrix is anything but linear. Understanding military theories does not necessarily result because of study from first principles; the model that philosophy leads to theory, which in turn leads to strategy, is an abstraction. It may be that only by pondering a strategy might an underlying theory be constructed and that only thereafter is the supporting philosophy uncovered. In other words, a lifetime of practical experience may lead someone to look backward from tactical knowledge to discern a strategy, then a theory and then a philosophy. Thus, although professional knowledge is critical for understanding, it would be a mistake to believe that such knowledge can only be acquired formally and in a linear progression. For the great majority of military professionals, the reality is likely the opposite: practical experience or the observation of certain tactics and strategies – as with Jomini studying the successes of Napoleon – may be the initiating action.

Whether progressing from formal education or from practical skills, the need for military professionals to study war and theory should be self-evident. Ironically this need is greater in small forces than in large ones. Joseph Stalin reputedly once quipped that quantity had a quality all its own. In other words, with greater resources to expend you can afford to be less professional. The reverse is true with small armed forces. They do not have the luxury of being able to make mistakes that waste resources. They must rely on the ability and professionalism of their leaders. This reliance allows the smaller forces to maximize the effects of their limited resources. Still, this professionalism is reflected in more than mere skills. It is also reflected in the quest for creativity in seeking unorthodox solutions.

In an article in the *Royal Swedish Academy of War Studies Professional Journal*, T. Jeppsson and T. E. Walther point out that to fulfill their mandate from society, military leaders have a professional obligation to have both a deep and a broad understanding of their profession. This fact is especially true for small nations:

> The true professionalism of military leaders reflects their ability in combat

THE UTILITY OF THEORY

skills as well as their creativity when resorting to unorthodox solutions. If the officers' knowledge of the theoretical framework is weak, there will be less understanding of the needs to change obsolete structures ... Within small nations and their armed forces, human related skills and qualities are of paramount importance. A high level of training, leaders at every level with the ability to lead possessing operational and tactical skills which reflect flexibility and ability to use unconventional methods, become absolutely necessary qualities. We have known this throughout history, but we need to be reminded that military theory and military history both still play an important role in creating independent-minded commanders and in setting up successful military organizations.[4]

But how should this analysis proceed? Is war an art or a science? Is it both? These questions comprise one of the most enduring controversies in PME. Cogent arguments exist on both sides. War continues to demonstrate both characteristics in its many manifestations. War is not unique in this respect; other aspects of human experience share this duality. For instance, there is an analogous problem in physics: Is the nature of light best understood to be particulate or wave-like? The straightforward solution is to admit that it is both. Light behaves simultaneously as particles and as waves. Although this is not entirely satisfactory, it does allow the nature of light to be investigated without being hampered by an incomplete theory. So, too, with war. War has a dual nature, an internal dichotomy. It can be described as an art, or as a science, or as both. In attempting to understand this dichotomy many military writers have resorted to the ancient Confucian concept of the yin and yang. In Chinese philosophies like Confucianism and Taoism, everything contains, as an integral part of its nature, a small amount of its opposite. Thus, the art of war contains elements of science and vice versa.

Although not definitive, the table below offers some insight. It is illustrative of observable tendencies. There is no attempt here to settle an ancient quarrel. It is sufficient to be aware of this dichotomy and to appreciate that this argument is one of the several long unbreakable threads that is woven through the intellectual history of war, the threads that comprise the fabric of the matrix.

We are again reminded of the six blind men of Indostan and their perceptions of the elephant. To the foot soldier in a trench or the subaltern in a small engagement, war may be pure science: the ranges of the weapons, the size of the battlespace, the numbers of troops, the amount of fuel, the amount of ammunition. To the commanding general, operating from a political directive that may be both broad and vague, she too must contend with the science of war as did Erwin Rommel, for instance whose orders from Hitler were simply to go to Africa and occupy as many British forces as possible. More likely, however, the senior commander will face the art of war: the personalities of political leaders or those of his subordinate commanders, the morale of the formation,

25

STRATEGIA

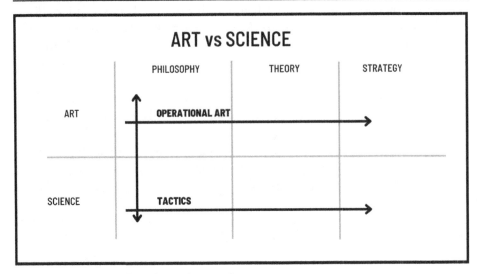

or the ideology or policy that is driving the campaign.

There is ample empirical evidence that at the higher levels war has a greater tendency to exhibit the appearance of an art whereas at the lower tactical levels it manifests itself more as a science. This is by no means a rule, but when you study great military commanders it becomes apparent that some have had insights that others lacked, that a few have even demonstrated genius in unexpected ways. The analogy with painting begs to be made; anyone can apply paint to canvas, few can be considered artists. Likewise, whether investigating the artistic or the scientific nature of war, there is a need for sound theories built upon solid understandings of one or more philosophies.

Frequently, as theorists like Clausewitz, Jomini, Mahan and Douhet offered new interpretations and insights into war's nature, new schools of thought grew up around them and their ideas. Not surprisingly since Clausewitz had been an original member of the Prussian Army Reform Commission in 1807 and went on to become the Commandant of the *Kriegsakademie,* the German army adopted him. The US army, having been influenced by the French beginning with their revolution and continuing through West Point instructors like Henry Halleck, adopted Jomini. The US Navy adopted their own Alfred Mahan. Almost all large air forces, the US Air Force, Royal Air Force, and all Commonwealth air forces, adopted Giulio Douhet. The 'discovery' in the West of Sun Tzu has seen his writings adopted by most NATO forces. Each of these adoptions led to the formation of a school of thought centred on the thoughts of that individual. The ideas espoused by these schools thereafter became crystallized and institutionalized as national doctrines. Thus, Clausewitzian thought spawned German doctrine, which resulted in the 'German Way of War.' Jominian thought did the same for France and the US – although it should be noted that the two are not the same. Therefore, the question is: What difficulties arise when the casual use of terminology and the borrowing of doctrine lead to the syncretic mixture of military schools of

THE UTILITY OF THEORY

thought?

It is unrealistic to expect that one school of thought could exist in complete isolation from all others. The cross-pollination of opinions and ideas is quite common, natural, and often extremely productive as with the Hegelian process of thesis-antithesis-synthesis. But unexpected difficulties can arise when ideas that are inconsistent with each other are artificially combined. Canada, with its practice of sending senior officers to foreign colleges and command billets is a prime example. These inconsistencies can result from many factors. Sometimes it is because the advocate simply does not understand the difference between one term and another. Sometimes it is because of the infelicities of language; sometimes it is the incorrect understanding of a foreign term.

The use and abuse of military English is elegantly described by Colonel Antulio J. Echevarria II in his essay "Towards an American Way of War". Echevarria is critical of Russell F. Weigley and his well-known treatise, *The American Way of War* for Weigley's incorrect labelling of various strategies, conflating attrition, exhaustion, and erosion, for example. Whatever the reasons, such mixing, borrowing and intermixture can have serious implications regarding the understanding that leaders, both uniformed and civilian, have of the nature of war. American English is notorious for its mixing and matching of terminologies. This issue becomes even worse when a term is taken from another language. Americans are not renowned for their precision with language and many problems of understanding have at their heart this inexactness of language. At least part of Brian Linn's essay "The American Way of War Revisited", which also rails against Weigley, in the *Journal of Military History*, stems from how the German concepts of 'attrition' and 'annihilation' have been mistranslated into English. Another example of this effect can be found in American army officer Robert Leonhard's *The Art of Manoever*. Leonhard takes Clausewitz's term *Schwerpunkt* and morphs its meaning into something completely new!

Consider the creation of military doctrine. If the foundations of doctrine are flawed, what flows from it will be predisposed to failure. The American army case is instructive. After Napoleon's final defeat, interest grew in the interpretations of his success. Jomini became the most widely read of Napoleon's interpreters and his influence spread quickly, eventually forming the basis of American doctrine. Jomini's theories of internal lines of communications and lines of force were very popular to the technically oriented engineers of the fledgling US army, especially at the newly formed academy at West Point. Over time, his theories took on the mantle of revealed truth. The US army did not appreciate that Jomini was a *theorist*. He was accepted as a *strategist* and his ideas slowly permeated all aspects of American military thinking. Of itself, this influence was not a bad thing. But a century later, when the US army discovered Clausewitz, they again misunderstood that he, too, was a *theorist*. Clausewitz's theories began to work their way into American doctrine. To this day, US army doctrine has an internal inconsistency because it simultaneously espouses the theories of Jomini

27

STRATEGIA

(*war has immutable principles*) and those of Clausewitz (*there are no principles of war*). American professional journals still contain regular, heated debates that attempt to rationalize these two opposing theorists. Add to this confusion what several writers, including Henry Kissinger, have argued, that as a society, America seems to regard theorizing in general as a futile intellectual exercise. Further, the US forces recently incorporated the works of Sun Tzu, a *philosophy* that is often misunderstood as a *strategy*. The risk of confusion is great:

> Some studies on the philosophy of war included views on the foundations of strategy (Fuller's classical study on the foundations of the science of war) and certain strategic studies, such as Liddell Hart's *Strategy: The Indirect Approach* contained a profound philosophical or socio-political component. This is also true of the recently published studies by André Beaufre and collective works on contemporary strategy. The two disciplines have thus often merged or one of them has absorbed the other.[5]

The problem grows. Without a clear understanding of the differences among philosophy, theory and strategy, issues become muddled. Understanding and using philosophical, theoretical, and strategic terms and concepts correctly are not difficult, but those who use these terms incorrectly or indiscriminately do other members of the community, especially those in uniform, a great disservice. Some sort of universal model or standard would help. Military experts must appreciate that their words are weapons and that their influence is potentially lethal. This is particularly true in Western democracies, where complete military novices may find themselves as presidents, prime ministers, or ministers of defence. Demands for precision of language should not be dismissed as mere sophistry; there are real implications for both the uniformed military community and its civilian leadership.

Military officers are not the only ones who are guilty. Academics fall victim to the same errors. In the renowned anthology *The Makers of Modern Strategy*, the editors collected a series of essays on military strategy from the early Middle Ages to the modern era. This collection by leading authorities of the time has become a benchmark in the scholarship of both strategy and theory and remains a standard text not only at military colleges but also in universities. Those studied range from Machiavelli to Napoleon to Helmuth von Moltke to Bernard Brodie. However, the very title is misleading. Although all the individuals discussed had important impacts upon war, not all of them were strategists. Some have had little or nothing to do with strategy. Likewise, Michael I. Handel in his book *Masters of War* compares the writings of Clausewitz, Sun Tzu, Jomini and Machiavelli. Handel's book is authoritative, and many respects, it is brilliant. Yet, many of the work's implications are misleading. Handel relates his subjects' mastery of war to strategy but was any of them a strategist? No. None of them.

THE UTILITY OF THEORY

Professor Handel is in good company. Many eminent scholars from Sir Michael Howard to Bernard Brodie also refer to Sun Tzu, Clausewitz, and Machiavelli as strategists. In his Taoist treatise *The Art of War*, Sun Tzu exhorts the reader to understand the nature of war, the nature of man and the interplay between the two in the natural world. He explains, for instance that water shapes its course according to the nature of the ground over which it flows and so a soldier shapes his victory in relation to the foe he faces. This is philosophy as surely as are the writings of Confucius. Conversely, Clausewitz, in *On War*, presents a military theory. His theory attempts to explain the overwhelming success enjoyed by Napoleon just as Jomini's contribution to military thinking in his *Précis* was his attempt to unravel the mystery of Napoleon's success.

Clausewitz and Jomini were not strategists; they were theorists. Jomini, did not establish a system of beliefs; he created a set of rules or principles. Certainly, their ideas have influenced strategy – Clausewitz's distinction between *Absolute* War and *Actual* War or Jomini's concept of interior lines have become so widespread and so embedded in modern doctrine as to be invisible –they are so commonplace that their roots in *On War* or *Précis de l'art de la Guerre* are no longer obvious – but they are theoretical constructs, not strategy. What about Machiavelli? This Florentine civil servant came closest to being a strategist. His most famous book, *Il Principe*, was a veritable do-it-yourself guide on how to gain and hold political power. Nevertheless, Machiavelli's real contribution to military art and science was not his pamphlet but his philosophy of power, including the military, and how it should be used in a political state. For this reason, he is considered by many to be not only the father of modern warfare but also the father of political science. Thus, he was neither strategist nor theorist; he was a philosopher. If such learned professors can fall victim to errors of mislabelling, is it any wonder that confusion reigns?

Summary and Conclusions

The study of Military Theory suffers from many problems: Language, both generally and specifically by the greater military community, is imprecisely and loosely used. Although widely studied, there is a lack of an internationally accepted model of what components comprise Military Theory. There remains confusion among the meanings of philosophy, theory, and strategy; and the question of whether war belongs more correctly in the domain of art or science remains moot.

The inaccuracy of language is a human failing that will likely remain so long as humankind continues to use different languages. This issue is best left to sociologists, although a little rigour within professional military circles would go a long way towards ameliorating the situation. The Bundeswehr tries in this respect, whereby General Staff officers are admonished to be exact in their communications and use only so-called *Generalstabsdeutsch* as opposed to *Truppendeutsch*. The problem is also cultural since

English, which is the West's least precise language is also its most widely used and we should keep in mind George Bernard Shaw's humorous adage that America and Britain are nations separated by a common language. In any case, the imprecision of language is to some extent as much a symptom of widespread misunderstanding as it is a problem.

The lack of an agreed upon international model of Military Theory is perhaps more vexing. The difficulty lies in the fact that unlike the 'hard' sciences, militaries and their theories of war lie within national, or a cultural, domains. But there is hope. In the same way that early scientific theories tended to be either parochial or nationally proprietary so, too, are military theories today. (Electricity used to be considered to flow positive to negative in England but negative to positive in North America). Nonetheless, with the advent of NATO and the growth of globalization, the potential exists for all Western militaries to eventually come together to create a uniform, international model. That would be a good start.

The disturbing tendency for the military community to mix and match ideas from different theories – sometimes in opposition to each other – to form doctrine that is fraught with internal inconsistencies is a problem that is likely insoluble, at least in the multi-cultural militaries of the West. Education will certainly help but a long-term solution remains doubtful. Thus, despite the many barriers, the study of Military Theory is not just worthwhile, it is necessary – particularly for smaller nations. If Military Theory is to be 'taken back' by military professionals and not left to sociologists, psychologists, and anthropologists, then continued study and education by uniformed personnel is necessary. Only through continuous study can progress be made towards a universal understanding of the nature of war.

Accepting these shortcomings and that study is therefore poorly focused, it is reasonable ask: Why study it at all? Will it assist military leaders in the performance of their daily duties? Will this study make war less likely? Will it make the world a better place? The answer is yes – to all the above. It is a truism that soldiers who have experienced war are the least likely to recommend it as a tool for conflict resolution. The more society understands war, the less likely it is to occur. The ultimate utility of studying war is that it will make life better because only when humanity better understands the nature of war, will be in a better position to improve the lives of all peoples.

Theory helps to explain life. In the same way that unknown technology takes on the appearance of magic, the absence of theory would make all events either unpredictable or unexplainable. Without theory, all human action can be perceived and explained as mere happenstance. The human mind requires order to clarify what it observes. Harvard professor Samuel Huntington offers a cogent summary:

> Understanding requires theory; theory requires abstraction; and abstraction requires the simplification and ordering of reality. No theory can explain all the

THE UTILITY OF THEORY

facts ... the real world is one of blends, irrationalities, and incongruities: actual personalities, institutions, and beliefs do not fit into neat logical categories. Yet neat logical categories are necessary if man is to think profitably about the real world in which he lives and to derive from its lessons for broader application and use.⁶

Although Huntington was referring to politics, the same is true of war.

Notes

1. Gérard Chaliand, *Anthologie Mondiale de la Guerre Stratégiques des Origines au Nucléaires*, (Paris: Robert Laffont, 1990), xv-xvi. (My translation).
2. Williamson Murray, "War, Theory, Clausewitz and Thucydides: The Game May Change But the Rules Remain," Marine Corps Gazette, (January 1997), 62.
3. Antoine-Henri de Jomini, *Précis de l'art de la guerre*, (Paris: Éditions Ivrea, 1994), 365. (My translation).
4. Tommy Jeppsson, and Tor Egil Walther, "Military History and Theory as a Professional Foundation," Royal Swedish Academy of War Studies Professional Journal, (Stockholm, 2001), 62.
5. Julian Lider, *Military Theory: Concept, Structure, Problems*, (Aldershot: Gower, 1983), 380-381.
6. Samuel Huntington, *The Soldier and the State: The Theory and Politics of Civil-Military Relations*, (Cambridge, MA: Harvard University Press, 1957), 1.

" What a society gets in its armed services is exactly what it asks for, no more, no less. What it asks for tends to be a reflection of what it is. When a country looks at its fighting forces, it is looking in a mirror; the mirror is a true one, and the face that it sees will be its own.

General Sir John Hackett
The Profession of Arms

MILITARY THEORY AS SOCIAL HISTORY

Introduction

MILITARY THEORY IS NEITHER MONOLITHIC NOR HOMOGENEOUS. Nor is it universal. It is rarely uniform across national boundaries and cultures. Views of war change with time and place. Different countries subscribe to different military theories, even when those countries are close neighbours as in the case of France and Germany. Military Theory contains a significant social component and comprehending this component may help to explain why humanity has made so little progress in understanding war's true nature. Therefore, we turn our investigation to the social components of military thought and theory as well as the linkages among society, history, and military theories. We will consider the concept that, historically, ideas and not technology drive innovation. We will review history's influence upon culture, theories, and military doctrine. We will look at how some cultures borrow military concepts from each other, and we will consider the uniqueness of each military culture, using a short case study to highlight the consequences of accepting a foreign military paradigm without fully understanding its roots. By comparing societies and their military cultures, as well as some of their theories, we will gain a better appreciation of some of the links and influences that form the Complex Matrix we seek to understand.

Why are military cultures and the military theories that they espouse so varied? Why does American military theory differ from Russian military theory? Why do both diverge from those of France, or Germany, or Great Britain? If we can agree that war is a universal human endeavour, then why are there no universally accepted principles that govern its conduct? Why does the theoretical vision of war change from country to country, from culture to culture and from century to century? We can accept that theories change over time, but what causes them to change as they cross international boundaries? The range of influences spans from none to a broad heterogeneous

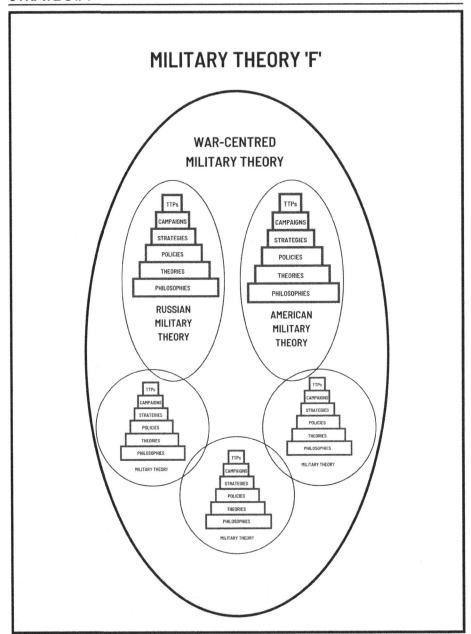

mix. The US army's keystone doctrinal manual, Field Manual (FM) 3-0, is based on Clausewitz but with a strong influence from Jomini. The Bundeswehr equivalent manual, HDV 100/100, *Allgemeine Führungsgrundsätze*, is pure Clausewitz. The equivalent Canadian army military manual, (CFP) 300, is a typically Canadian pastiche; it is part Clausewitz, part Sun Tzu, and part Jomini, with influences from generals Andrew McNaughton, Guy Simmonds, Erich von Manstein and Bernard Law Montgomery!

Society and Military Theory

The Complex Matrix inextricably links history, geography, social norms, culture, and military theories. As components of that matrix, they all play influential, if indistinct and varying, roles. Nevertheless, for the purposes of our investigation, knowing exactly how and why these links influence each other is less important than appreciating that they exist, and that history, geography, social norms, and culture all inform military cultures and military theories. Thus, appreciating that these theories frequently do not survive intact the crossing of borders, the models presented previously at [Military Theory 'D'] expands once more to become the model at left as [Military Theory 'F']. The diagram is purely representative; there is no attempt to show all possibilities. Nonetheless, it demonstrates that there is no universality or exclusivity in the greater domain of Military Theory. Further, it shows that states and cultures have their own theories that may include pieces of others' theories. Lastly, although for simplicity's sake not shown in the diagram, all these characteristics hold true for both the art and the science of war.

An understanding of Venn diagrams is useful to appreciate the model. What the diagram demonstrates is that within the universal domain of Military Theory there are many, many military theories. These *lesser* theories are derivative of their national cultures. In some cases, they borrow and lend ideas to other countries. For instance, the small oval below the oval marked Russian Military Theory might be Hungary or Poland and likewise under the American oval might be Canada and Australia. The small oval in the bottom centre represents a country that is highly independent of all others, a country such as Costa Rica or Iceland.

The science of war has clearly been a story of continual technological change. Beginning with clubs and stones, humankind has altered how it fights to the point where commanders can now launch weapons on one continent to strike targets thousands of kilometres away with deadly precision. Unclassified data regarding cruise missiles indicate that even across thousands of kilometers of range, the missiles can strike their intended targets within meters of accuracy. It is an astounding technical feat but focusing upon this technical achievement can lead to false lessons. It is a widely held misconception, for instance, that technology has driven the evolution of war. Although true that the history of warfare has been a progressive search for technological advantage, that is not the same as saying that technology has driven the evolution of warfare. For instance, weapons have progressively developed greater range, but this is because there was a need for this greater range, not because it was technically possible. Occasionally a technology is developed and thereafter an application is sought, but in most cases, a need is determined, and a technological change is made to

STRATEGIA

address that need.

Of course, it is far easier to demonstrate the leap from sling shot to ballistic missile than the leap from command and control of a hoplite phalanx to the concept of *Auftragstaktik*. In the twenty first century, the wonders of weapons' development and new technology are a popular view of the changing face of war. The two Gulf Wars saw the popular media obsessed with technology. Viewers who had never heard of a SCUD missile saw pictures of them on practically every newscast. Whether you listened to ABC, CBC, CBS, or CNN the anchors and the retired military officers, who acted as analysts, always returned to the same issue: The coalition had access to heretofore unheard-of technology. Whether this technology was in the form of "smart bombs" or Tomahawk missiles that could find targets from 1,000 kilometres away, or tanks that could see through smoke, the dominant theme of the discussion seemed perpetually to be technology rather than idea based. Rarely did you hear of anyone wondering if the enemy had a coherent strategy or doctrine that was built on sound military theory. The exception that stands out was General Norman Schwarzkopf in the 1990 war. In his cessation of hostilities briefing, he derided the Iraqi armed forces for not having officers who were trained in theory or a coherent command and control doctrine. But General Schwarzkopf's brief comments were unusual.

Technology, however fantastic, has never been the formative engine of change; the popular view is both facile and inaccurate. The development of new weaponry has never been the principal driving force of military innovation. In fact, the opposite has been true; technology has been the servant of ideas. The single most important influence upon military change has always been the intellectual mastery of new concepts of war. In other words, it has been the interpretation and reinterpretation of Military Theory. As Basil Liddell-Hart put it his book *Strategy,* the crucial factor was "the influence of mind upon mind – the most influential factor in human history." This truth has been neither widely accepted nor well understood. Notably, this has also been the case among professional soldiers. Often, the lure of new weaponry has been too attractive to resist; the belief in the ability of technology to overcome a tactical problem has been too strong. A society wedded to the belief that technology drives human advancement – as is

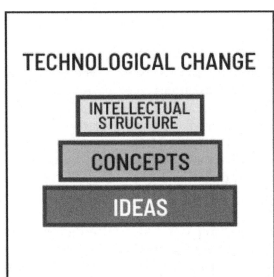

modern Western society – can easily be blind to the fact that this belief puts the cart before the horse. Ideas drive human development and always have. How those ideas are developed and then manifested is the reality of humankind's development. This argument is not complicated: ideas create concepts, concepts create intellectual structures and intellectual structures drive technological change. It is almost Cartesian in its simplicity. Humanity drives civilization, not an accidental discovery of a better mousetrap. The parallel with the model presented previously as [Military Theory 'A'] is obvious:

Ideas must come from somewhere. They are not self-generating, not *sui generis*. The wellspring of humankind's ideas, as well as the force that drives their development, is human culture. The more accepting of new ideas a society is, the more ideas it will generate. It is not a coincidence that free and open liberal democracies have been the source of so many inventions and developments. It is a straightforward connection: Societies that value new ideas, whatever their belief systems, create more of them. Whether we are discussing the Classic Greek Age, the Islamic Golden Age or the European Renaissance, the model holds true. Where knowledge is valued, progress is the result.

In the 20th century, a strong military example presents itself. The unexpected military successes of the Wehrmacht early in the Second World War were not due to Germany's technological superiority. As advanced as the weapons of Krupp, Thyssen, Messerschmidt, *et al* may have been, in some cases, Germany had technically inferior weaponry to its enemies and still won. German victories were invariably due to superior intellectual concepts and better doctrine. This point of view has been supported by respected military historians from Michael Howard to John Keegan to Trevor Dupuy, to Robert Citino. During the invasion of France in 1940 the Germans had both inferior weaponry *and* inferior numbers. Their advantage was in their military theories, and doctrine which encompassed flexible conceptual models, educated leadership, and most particularly, innovative tactics: "It was the panzer division, not the panzer, that defeated the Allied armies. The German commanders created mission oriented, customized groupings that reflected a sound doctrine and an experienced organization to drive it."[1] Military culture and experience, not technology, were the winning advantages that Germany had over its multiple enemies.

The real Wehrmacht advantage was in having a military culture that was deeply imbued with the practice of adapting new ideas and allowing them to alter the military intellectual structure. This mindset is borne out by multiple historians of the subject like Robert Citino, in *The Path to Blitzkrieg,* Jehuda L. Wallach, in *The Dogma of the Battle of Annihilation*, and Martin Kitchen, in "The Traditions of German Strategic Thought," in *The International History Review.* As Williamson Murray explains, the German military's culture of incorporating the best and applying it to an existing broad and flexible ideological paradigm served them throughout the Second World War:

> German doctrine allowed them to innovate successfully in armored warfare. In contrast the French never had time to incorporate innovations on the battlefield because their collapse came so swiftly; the British, who possessed the time due to geography, never displayed the talent. Their defeats in North Africa reflected a lack of a doctrinal framework in both planning and execution. Their officers were not capable of adapting to a larger framework of war. They remained prisoners of the compartmentalized conceptions of warfare that characterized their branches. They remained gunners and infantrymen to the end ...[2]

Thus, new ideas *used* new technology to alter an intellectual framework or paradigm and not *vice versa*.

The influential Berlin professor and historian Hans Delbrück held that societies were the children of their history, '*Gesellschaften sind die Kinder ihrer Geschichte.*' In other words, the past shapes the future; it is pure cause and effect. Delbrück, who died in 1929, became one of the German General Staff's most widely read historians but during the First World War, he became one of its most vocal critics. He was the first to solidly link warfare to politics and economics. His concept of *Fachgescichte*, or specialty studies, became an essential tool of the staff. His analyses of ancient battles, and of the Battle of Cannae in particular, became a cornerstone of German army doctrine and remains so to the present day. Delbrück argued that the internal discipline in Roman society allowed the Romans to create their highly disciplined legions, which in turn allowed them to use the tactics, which they did with such devastating effect.

Conceptual Grafting

Like all societies, Canadian society is unique because of its past. American society is likewise so and although the two societies share many similarities, as well as a great deal of history, they remain distinct. Societies display certain characteristics based on how they have experienced the past. Despite the attempts of many to try, no country can escape its own history. To deny the influence of the past is nonsense. Still, it does occur and nowhere is this denial more poignant than in modern Germany. It has struggled since 1945 with what has been dubbed *die unbewältigte Vergangenheit*, or the 'undigested past'. The assertion by some postwar German historians that the twelve years of Nazi rule, from 1933 to 1945, were an interregnum or *Abbruch* in German history has been hotly debated for decades. The controversy is known in Germany as the *Historikerstreit* or 'historians' conflict'. Fritz Fischer initiated this controversy in 1961. Fischer was trained in the tradition of Wilhelm von Humboldt, Johann Gottfried Herder, Georg Wilhelm Hegel, and Leopold von Ranke. Before the Nazis rose to power in 1933, historians were the political mentors of the nation. Even though many non-German historians viewed the ascent of Nazism as a natural continuance of the country's historical development, this was not widely accepted within Germany. Fischer's research convinced him of the truth of the foreign view. He shook the national confidence. He believed Germany and the Kaiser to be directly responsible for the World War I. He supported A.J.P. Taylor's assertion that the absence of a successful and lasting liberal middle-class revolution in 1848, as had occurred in France and Britain, and the subsequent unification of the German Reich under Prussia, created fertile soil for the growth of political dictatorship.

Consequently, it may be politically correct in today's *Bundesrepublik* to insist that the Austrian Adolf Hitler and his gang of murderous thugs somehow 'stole' Germany from the peace-loving Germans; but the historical truth is otherwise. Hitler and his party were elected and supported by the German people. Germany, therefore, like France, Japan, Italy, Spain, Guatemala *et al*, which have had military organizations interfere with their civil governments, remains a prisoner of its past. This is a straightforward and inescapable fact. Professor Delbrück was correct, and his words act as a warning to all societies: Where you come from largely determines where you will go. This connection with the past has a strong influence upon countries' military thought and theory. All nations shape their cultures through their histories, and subsequently their thinking. The Italian General Staff is a good case in point. During the 1920's and 1930's, at the same time as the German Troop Office was experimenting with mechanized formations and the effects of airpower, Italian military culture put great store in a General Staff officer's equitation and fencing skills. Even though the father of air power theory was the Italian Giulio Douhet, and the army was an early proponent of mechanized warfare, the culture of the General Staff insisted that gentlemen ride and fence and that staff work was somehow an inbred talent. Italy's battlefield performance during World War II are indicative of the results that this belief fostered.

Wherever countries borrow military theories, the borrower nations are usually obliged to make modifications before they can successfully apply them. All NATO armies have drawn upon German doctrine. Some, like Denmark, simply reprint most of the doctrine, holus bolus. Looking through any of the doctrine manuals like the Canadian CFP 300 or the American FM 100, a German *Führungsakademie* graduate will look in vain for a four-step Mission Analysis. The American version can be anywhere from 12 to 29 steps. The Canadian version, which seems constantly to be changing, and with little or no recognition of the process being borrowed, has been anywhere from 8 to 24! But borrowers must respect the unique historical and cultural components of every country's military theory; where they ignore them, glaring inconsistencies can result. Thus, whenever the borrower does not take heed of history and culture, the transplantation of military theories inevitably fails.

In the late twentieth century, NATO armies became enamoured of what had been an almost uniquely German view of war. The German school had been forged in the fires of the French Revolution and annealed during the Wars of Unification. It was born of struggle, desperation and in large measure from the Prussian legacy left by Frederick the Great. To understand the errors made by NATO in espousing the German school we need to look briefly at how it came to be developed; we need to look to the birth of *Auftragstaktik* or what Anglo-Saxon militaries have come to call Mission Command.

The social and political cataclysm of the French Revolution changed forever how Western armies fought. New philosophies spawned new theories that led to new policies and strategies. These strategies then created and applied new tactics, techniques, and procedures. The TTPs became tools in waiting for someone with the skill to combine them into a new form of warfare. This someone was Napoleon Bonaparte. Napoleon can be credited with many things, but from the standpoint of Military Theory, the most important was his forcing of a fundamental change in Prussia and its army when he destroyed it at Jena in 1806. He obliged the Prussians to completely re-evaluate how they perceived war. The army was broken. Civil and social structures were shaken. Political disruption, military culture and national survival coalesced to create a tool, *Auftragstaktik*, unique in Western history. The creation of *Auftragstaktik* is a case study in military theory as social history.

The watershed upheaval of French Revolution was multifaceted. Europe's largest monarchy collapsed under the weight of its own incompetence therewith threatening to take down the neighbouring monarchies as well as the entire political structure of Europe. This collapse forced complete transformation upon Europe. The *levées en masse* of the French army brought heretofore-unseen numbers of troops to the battlefield while Napoleon's uncanny ability to bring greatly dispersed forces together into single pitched, almost cataclysmic, battles were unknown in modern times. In doing so, he achieved what in modern parlance would be described as 'battlespace dominance.' This dominance led to "strategically focused, sequential operations and engagements culminating in a dominating maneuver to destroy the enemy's armed might."[3]

Observers watched his consummate military skill and were awestruck. The result was that military operations *per se* took on an overwhelming importance. Campaigns and battles came to be seen as prime determiners of a nation's fate. Politics temporarily took a back seat to combat-based international relations. "Military operations were no longer one part in a complex counterpoint of international negotiation: they played a dominant solo role, with diplomacy providing only a faint apologetic obligato in the background."[4] Napoleonic strategy confined the rigid, linear warfare epitomized by Frederick the Great's Guards Regiment to the dustbin. The use of *tirailleurs*, or skirmishers, called for a new understanding of war. It called for a shift in how leaders, at all levels, perceived battle. Slow, precise, geometric formations, which were highly drilled and brutally disciplined, were facing the hit and run tactics of small groups. Light infantry skirmishers were being trained to think for themselves, and to use ground cover to best effect. Mindless drill was forced to give way to flexibility of both thought and action. Although previously discussed in the military literature, no significant change in how armies fought appeared until the French shattered their ancient monarchical system and put a nation of politically inspired peasants under arms. The new philosophies of equality and the worth of the individual led to the massive armies created by *levées en masse* and to the opening of officers' commissions to non-aristocrats. These two changes combined to cause a dramatic alteration in warfare. The effect was synergistic: social change forced military theory, strategy, tactics, and leadership all to transform in concert with each other.

Many officers in the German states watched these changes diligently. In 1801, an artillery major gave up his Hanoverian commission to become an officer in the Prussian army. He was immediately posted to the *Generalquartiermeisterstab* in Berlin and in July of that year joined the *Militärische Gesellschaft*, or Military Society, an officers' association, which shared tactical studies and lobbied for reform. In fact, he was asked to join and at his first meeting was named by the society as its director. That major was Scharnhorst and officers who would later reform the Prussian army and thereafter influence the Western understanding of the nature of war flocked to the society. Among them were lieutenants Carl von Clausewitz and Rühle von Lilienstern. "It is significant that these officers came ... not from the ranks of the rooted Pomeranian Junkers but from other regions and other sections of society."[5] In the *Gesellschaft*, discussions focused on the new French style of warfare and whether the Prussian army could incorporate these new tactics. Further, the philosophies of the German Enlightenment, or *Aufklärung*, were influencing the *Gesellschaft*: some called for increased freedom of action at lower ranks, for the empowering of the individual as well as the recognition of the importance of the moral aspect of war. These ideas had a profound impact upon the young Clausewitz and the early influence of this intellectual intercourse upon him cannot be overstated.

The reformers did not live in a vacuum. They knew that armies were living organizations that needed nurturing. But theirs had crumbled in defeat. They saw that the Prussian army, once the envy of Europe, was now incapable of Napoleonic warfare. Prussia's defeat in 1806 brought the proof to the Prussian king's doorstep. Napoleon had crushed Prussia's once famous and feared army. Prussia was humiliated. As a matter of survival, the reformers adopted new and radical theories that grew to become part of the army's culture. Of these, the most successful became *Auftragstaktik*. Over the next century, *Auftragstaktik* grew to become symbolic of German military excellence and eventually – even after another series of defeats – a model for NATO to emulate.

In the 1980s, the British army rediscovered the operational level of war as well as the 'Operational Art' or *Operative Kunst* as the German army coined it. The term describes the abilities of commanders to function at the operational level of war and it is noteworthy that the Germans considered war at the operational level to be an art rather than a science. The British began to study some of the German commanders of the First and Second World Wars and subsequently became the first army in NATO to consider seriously the adoption of some German military theory relating to independent action and *Auftragstaktik*. This study occurred while the British army still held fast to unwavering obedience to superior commanders and strict adherence to positional defence:

> The best type of defence is when the front, and any vital ground can be covered with mutually supporting fire positions and obstacles, with further positions in depth, and with mobile reserves at hand to stiffen this framework when required. The enemy is defeated as he struggles to break into the defended area by concentrated direct fire from the mutually supporting positions combined with the best possible use of indirect fire resources. This concept is commonly known as positional defence.[6]

German warfare was, in many ways, antithetical to the British way of war. German military theory was heavily dependent upon manoeuvre and the primacy of the counterattack. It depended upon the willingness of subordinates to act independently and to use initiative. Adopting German military theory would mean that either the British army had to give up its own military theory, or it had to modify what was transplanted from Germany. In 1987, the German exchange instructor at the British Army Command and Staff College reported to his superior at the *Führungsakademie* that the British had shown a renewed interest in the battle techniques of the German armies of the First and Second World Wars. The German *Oberstleutnant* reported that although the students were eager to adopt the German doctrinal practices of *Auftragstaktik* and battlefield manoeuvre, the British Directing Staff were extremely reluctant to do likewise. The staff, more deeply steeped in the British tradition of positional warfare than the younger students, could not fathom how – or why – the British army, which had twice beaten the German army in battle, should adopt its doctrine.

Auftragstaktik, an almost mythical command function, was and remains the keystone of the German way of war. The basis of *Auftragstaktik* and the manoeuvre style of warfare was *Auswertung des Auftrages* or Mission Analysis. Using this four-

---

MISSION ANALYSIS

Step 1: What are my superior commander's intentions?
Step 2: What is the enemy's intent?
Step 3: What are my constraints, restraints and tasks, both assigned and implied?
Step 4: Has the situation changed fundamentally?

step process, shown below from HDV 100/100 *Allgemeine Führungsgrundsätze*, a military commander considered his or her part in the overall scheme of battle and established the mission. Subordinates reviewed their superiors' intent, considered the enemy intent, listed their tasks and finally, but most importantly, considered whether the situation had changed fundamentally. The process was continuous. It never stopped at any time and based on this ongoing analysis any subordinate could change a mission if the situation dictated. Deciding what fundamental change meant was always the purview of the individual commander – at all levels. Interestingly, HDV 100/100 makes no attempt to quantify or list what a fundamental change might be. Commanders, whether they are corporals commanding sections or lieutenant generals commanding army corps, are expected to exercise personal judgement in this respect.

This process had radical implications. It could result in subordinates giving up their assigned missions and creating new ones that might be diametrically opposed to their original orders. Such possibilities made British senior officers extremely uncomfortable. Their military paradigm did not allow such behaviour, so they modified the mission analysis process. Instead of having a subordinate choose a mission based on one of a series of tasks and how it fit into the greater scheme of battle, the modified method called for superiors to give their subordinates not only the series of tasks but also their mission statement. The need for mission analysis was thereby obviated. No mission analysis (at least by German standards), no *Auftragstaktik*. British soldiers could claim that they were adopting the German military theory, but they were doing nothing new. They were merely following orders as given. Nonetheless, the British manuals insisted that Mission Command had become integral part of the British way of war. This fundamentally German process, based on the philosophies of the *Aufklärung* and theories of generals Gneisenau, Clausewitz, Scharnhorst and Moltke did not survive transplantation into the British army. The British army had its roots in a completely different history, a different social philosophy, and a different military theory.

Although British doctrine stated something that appeared like what the Canadian army or the US army taught, its application was quite another thing. To interpret the doctrine correctly you had to read and understand all of Army Field Manual 1 and although extremely well written it was clear the British army has modified the German doctrine considerably in its application to make it work in their army. For instance, although a series of tasks was given to a subordinate and the subordinate was required to do a full mission analysis, the superior gave the subordinate a mission statement as a 'unifying concept'. What this did was have the subordinate go through the mission analysis process simply to clarify what he had been told to do. This was not bad, but it was not what the German doctrine called for and it took away the subordinate's freedom of action.

The British example was not unique. Many NATO countries imported and attempted to implant *Auftragstaktik*. Unfortunately, in one way or another, the British example was repeated. Professor William McAndrew, from the Canadian

STRATEGIA

Forces Directorate of History, liked to use an analogy to help Canadian officers understand the fallacy of this ideological transplantation: "No matter how carefully you transplant," he would say, "Cactus will not grow on the tundra – no matter how much you care for them." General Hackett was right. When a country looked at its armed forces it looked in a mirror. It was easy to overlook the fact that the people who interpret military theories were products of their societies. The military doctrine that armed forces created was a result of distinct philosophies and theories and could not help but be influenced by their cultures. For good or ill, all military theories embraced integral cultural components.

Since Military Theory has a cultural component and military cultures spring from their greater societies, these subordinate and smaller cultures usually display the prejudices, strengths, and weaknesses of their parent societies. This seemingly axiomatic statement is by no means without controversy. There have been occasions where militaries have isolated themselves from their parent societies and thereby subverted some of the beliefs and values of their parent societies. The late nineteenth and early twentieth centuries offer two distressing and related cases. The general staffs of Germany and Japan, for instance, isolated themselves from their societies. Because of the social power held by these military organizations, they became instrumental in leading their respective countries into war. Although extreme cases, they offer valuable lessons for all who study war and the civil-military relationship.

In hindsight, it is not surprising that the Japanese followed the same road as their German tutors. The modern Japanese General Staff was the product of a Prussian General Staff officer. The Japanese intentionally emulated the Prussian model. In 1885 a Japanese army delegation visited Germany and made a request of the Chief of the General Staff, General Moltke, to have a German officer to train Japanese officers in Japan. The Japanese wished to emulate Frederick the Great's use of the army as the 'school of the nation.' Moltke assigned a veteran of the Franco-Prussian War, Major Jacob Meckel. Meckel's tour of duty in Japan lasted only three years, during which he trained some sixty of the highest-ranking officers in tactics, strategy, and the organisation of the Prussian General Staff. His authority was so great that all other influences, British, French *et al*, were displaced. His teachings reinforced Japanese subservience to the Emperor by teaching his pupils that Prussian military success was a consequence of the unswerving loyalty to the king as supreme *Kriegsherr*. In the same way that the General Staff was linked directly to the Kaiser by an oath of personal fealty, in Japan the Samurai code of *Bushido* tied all military officers to the Emperor, even unto death.

National cultures and their militaries have their own internal laws, norms, taboos, hierarchies, and structures. It is this internal structure, which makes comparisons of military theories from different cultures so valuable. The British, Canadian, and American armed forces, for instance, share a common language. They have interlocking histories, have interoperable military theories and have been strong allies for more than

MILITARY THEORY AS SOCIAL HISTORY

a century. Nonetheless, each armed force is unique. That said, because of such close ties, the cross-cultural influences create interesting permutations and combinations. The Canadian military, having been a virtual clone of the British military for most of its existence is now most closely associated with the American military. Canadian uniqueness now lies to a great extent in being part British and part American.

Sun Tzu wrote about warfare from the perspective of his fifth century BC Chinese culture. Machiavelli did likewise from late medieval Florence. Clausewitz wrote as a product of the German *Aufklärung*. All, therefore, offer insights into the nature of war from distinctive viewpoints. In comparing these views, we may gain a better appreciation of the links between military thinking and social development. Professor Williamson Murray again:

> Military culture is shaped by national cultures as well as factors such as geography and historical experience that build a national military "style." The American military, for example, has always had to project its power over great distances. Even in the Civil War, which has exercised such great influence over the general military culture of the U.S. services, Union forces waged a war on a continental scale equivalent to the distance from Paris to Moscow. Germany, by comparison, was for centuries at the center of European wars, and consequently tended to neglect logistical problems.
>
> As with all human affairs, however, military culture is not immutable. Changes in leadership, professional military education, doctrinal preference, and technology all result in the evolution, for better or worse, of the culture of military institutions. The effects on culture, however, may not be evident for years or even decades, and may in fact be unintended consequences of other shifts.[7]

We cannot understand the military theory of a nation, therefore, without having at least some appreciation for its social culture in combination with its national military culture.

Consider the United States. The need to understand American military culture becomes particularly important now that it is the world's only 'hyperpower'. No other society has ever put so much trust in machinery or held so much faith in the power of technology. Americans have generally insisted that their military forces be closely connected to society at large. Hence, society's love of technology has been transferred to the US military and has subsequently had a profound influence upon US forces. "American military culture [has] historically emphasized scientific approaches to warfare to the point of holding an almost mystical belief in the power of technology to solve the challenges of war."[8] Clearly, it is important to understand this belief in the context of the greater society.

Such faith in technology has left the US military vulnerable to the preaching of futurists. Although less prevalent at this moment, in the last decades of the twentieth century, the husband-and-wife team of Alvin and Heidi Toffler held enormous sway within American government as well as the American military. Beginning in the 1970s with the bestseller *Future Shock* (1970), they followed up with more bestsellers like *The Third Wave* (1991) and *War and Anti-War*, (1995). The army and air force were particularly vulnerable to their influence regarding the importance of technology and its transformational ability. The immense popularity of the books gave the couple great credibility and influence. But not everyone shared this enthusiasm. Steven Metz, a US Army War College professor dismissed their predictions as publicity pieces. Others called their writing oversimplifications, psychobabble, vague clichés, and impenetrable jargon. Under such withering criticisms, much of their influence passed as quickly as it arose.

More recently, retired USMC Colonel Thomas X. Hammes claimed in *The Sling and the Stone* that war had entered a "Fourth Generation" of warfare or 4GW. The embrace was widespread but short-lived, and rightly so. The American reliance upon technology is not new, however. Neither is it uniquely American or all bad. England began the Industrial Revolution; through science, Germany became an industrial powerhouse; and as will be discussed later, the French Navy had a brief but ultimately disastrous love affair with technology at the onset of the twentieth century. Nevertheless, even when the belief in technology is initiated elsewhere (like Italy in the example below) the American military has inevitably taken the lead in that technology as well as in the belief that it would change the nature of war:

> Some ... warfare theorists have attempted to shield themselves from accusations of technological determinism by suggesting that we need not follow slavishly the technology wave. Rather, we should use our imaginations to determine what we want it to do for us and then develop the technology to fit those needs. This ignores the fact that technology development, much like the formulation of strategy and tactics, is a co-evolutionary process. New technologies emerge to either exploit or compensate for weaknesses in existing technologies. Inventing a [new] theory of ... warfare risks falling victim to the kinds of fallacies that Douhet encountered. Unable to see the future, he imagined one based on linear projections of extant technologies. Unable or unwilling to imagine counter-air defenses, or the limitations of strategic bombing in the face of a determined foe, he saw only a pristine view of air power, conducting operations with impunity against helpless, terror-stricken citizens.[9]

Douhet and his technology-based theories have long since been taken over by America. The love affair with technology by Western society has been building for hundreds of years. It stems from a European development that had its roots in the Enlightenment

MILITARY THEORY AS SOCIAL HISTORY

and the Industrial Revolution. The steady expansion of Europe's scientific knowledge, with the concomitant unspoken promise that technology could unlock the hidden secrets of the universe, fuelled this phenomenon. Although men such as Johannes Kepler and Galileo Galilei had previously made great discoveries, one individual stood apart from the crowd both in intellect and in influence: Sir Isaac Newton. His studies into light, matter, gravity, and the nature of the universe were the sparks that ignited the European fire of desire for more scientific knowledge. The introduction of his scientific method of investigation soon changed Western philosophy, mathematics, and warfare. Shortly after the publication of his *Philosophiae Naturalis Principia Mathematica* in 1687, Newton's theories began to permeate European society. His writings managed to codify and unify the work of all of those who had gone before him. His *Principia* quickly became if not the greatest scientific work of all time, certainly among the handful of most influential. The impact of Newtonian science upon the Western world has been almost unparalleled. The influence of Newtonian science on Western philosophy became so pervasive that it affected practically all facets of human life, including military theory. The ultimate effect of Newton's authority was that practically all fields of human endeavour adopted his scientific 'laws'.[10]

Newton's First Law of motion led to the conclusion that for every event there had been a previous event, which caused this outcome: cause and effect. This linkage soon became the basis for all reasoning. Any experimentally supported hypothesis had to demonstrate this cause-and-effect mechanism. Every sphere of human activity turned to science to solve problems, uncover principles, or lend authority to theses and hypotheses. Adam Smith in Economics, Thomas Malthus in population growth and Gregor Mendel in genetics all drew upon Newton's cause and effect methodology as did Charles Darwin in his theory of natural selection. The concept of cause and effect, discovered through Newton's scientific method, became the key to unlock virtually all aspects of man's existence.

Newton's ideas helped ignite both the Age of Reason and the Enlightenment. In France and England, the rising middle class, desperate to throw off the suppressive twin yokes of the Church and the nobility, embraced these movements. The rise of socialism in England, the *Risorgimento* in Italy and the *Vormärz* in Germany were all outgrowths of the Industrial Revolution and subsequent rise of the middle class. Science offered an alternative, secular, and more hopeful view of creation. The future could be liberated from an oppressive past. The ideas of the middle class suddenly gained places of prominence.

Armies and navies were not immune to the spread of this new thinking. If science could explain the workings of the physical universe, then could it not also explain the workings of war? If there were mathematical explanations for everything from the movement of the planets to the flow of human blood, to the inheritance of human traits, then perhaps science could uncover the 'eternal' principles of war:

The ideal of Newtonian science excited the military thinkers of the Enlightenment and gave rise to an ever-present yearning to infuse the study of war with the maximum mathematical precision and certainty possible... Indeed, the military thinkers of the Enlightenment maintained that the art of war was also susceptible to the systematic formulation, based on rules and principles of universal validity, which had been revealed in the campaigns of the great military leaders of history.[11]

But as armies and navies were embracing scientific rationalism and determinism, some social and political thinkers were already looking further afield. As some military thinkers were adapting the ideas of mathematical certainty to warfare. The intellectual tide was already beginning to turn. Indeed "this search for scientific certainty in military affairs was taking place at a time when thinkers concerned with other areas of human activity were beginning to question the whole idea of scientific certainty."[12] At about the same time, France underwent a cataclysmic revolution and Napoleon changed not only how nations fought wars but also how politicians saw them.

Politically, the Napoleonic Wars taught Europe a false lesson. War became the single most important aspect of political discourse. War had re-invented itself. Paul Schroeder, in his influential work *The Transformation of European Politics 1763-1848*, explained that the rules of international political discourse changed. Unfortunately, even as European soldier-scholars attempted to unlock the secrets of Napoleon's success by poring over reports of his latest victory, European statesmen were slow to notice that their roles had changed. The new model implied the suspension or interruption of politics, or even worse, that politics was now the servant of war. Once the model was accepted, it remained for a long time: "Nothing that happened in Europe during the next hundred years was to undermine the view that war now meant the interruption of political intercourse and the commitment of national destinies to huge armies whose function it was to seek each other out an clash in brief sanguinary and decisive battles."[13] The Western world would pay dearly in the next century for this misinterpretation.

After defeating Napoleon, the victors were determined to re-establish the social and political order *ante bellum*. The chief director of the peace, Austrian Prince Klemens von Metternich, constructed the 'Concert of Europe', a balance of power arrangement among the victorious European nations. The reactionary peace imposed across the continent by the Concert lasted more than four decades until the social revolutions of 1848 heralded the German and Italian wars of unification. These wars and those that that followed in the remainder of the century did nothing to change the common misperception of war. Further, the search for a climactic 'Napoleonic' victory in battle became *idée fixe*.

The Europeans almost completely ignored the American Civil War, with its four years of endless bloodletting. Europeans paid the price for this oversight in the horror of

MILITARY THEORY AS SOCIAL HISTORY

the First World War. Professional soldiers everywhere looked for the great Napoleonic victory that would cause the enemy to capitulate. But the quest for decisive victory through a single battle was in vain. Societies had changed. Technologies had changed. Battle had changed. The American experience had clearly demonstrated that modern technology had removed the self-sufficiency of armies in the field. The telegraphs, and railways had linked armies with their supply depots, with their political centres of gravity, and with an almost endless number of recruits. Defeated armies could be quickly re-supplied and re-equipped. Since the field commanders were no longer the political leaders, a tactical loss in battle did not automatically compel the political consequence of national surrender. The Battle of Gettysburg, 1-4 July 1863, epic by any standard, provided ample proof. Lee, having suffered a devastating defeat, withdrew from the field but the Confederate South went on to fight for two more years.

The North did not win the war with a single brilliant Napoleonic battle or even a single campaign. But it was certainly not for lack of trying. From 1861 to 1863, President Abraham Lincoln hired and fired a half dozen commanding generals, all of whom were expected to win such a climactic victory. Eventually the North won through a combination of the political determination of Lincoln, the long, tedious, often grinding campaigns of William Tecumseh Sherman and Ulysses S. Grant and an industrial base that was overwhelming relative to that of the South. The US Civil War had been the first war to be won by industrial might, by supply lines, and by logistics. It was also proof that the Napoleonic victories had been an anomaly. Students of military art and science may have been enamoured of the cavalry panache of J.E.B. Stuart, blinded by the tactical brilliance of James Longstreet, and awed by the operational genius of Robert E. Lee. But just as with the Wehrmacht almost a century later, battlefield proficiency did not translate into ultimate victory. Too many overlooked the political-strategic vision of the grimly determined Lincoln and the operational ruthlessness of the methodical Grant. The lesson of what some have dubbed the first industrial war was clear. Strategists may have admired the operational elegance and creativity of Stonewall Jackson or Lee, but it was the grinding attrition of both Grant and Sherman that gave the North its victory. Focusing too narrowly on only the military aspects of war missed the point, and as Professor Howard said, was a waste of time:

> There are few more tedious and less profitable occupations than to study the campaigns of the great European masters of war in isolation – Maurice of Orange, Gustavus Adolphus, Turenne, Montecuccoli, Saxe, even Marlborough and Frederick the Great; unless one first understands the diplomatic, the social and the economic context which gives them significance, and to which they contribute a necessary counterpoint.[14]

As Clausewitz had warned, the 'logic' of war is political, whatever its 'grammar'.

STRATEGIA

Summary and Conclusions

What has this glimpse at the connection between war and social history taught us? Studying war and Military Theory without studying their social and political components is counterproductive. Military Theory contains manifold linkages. Culture, history, and military theory are but a few of the components of the Matrix and there are critical social components that build and shape it. Despite being a universal human activity, Military Theory's constituent social dimensions transform it in time and space as well as across political borders. Any student of war who wishes to gain a better understanding of its nature must appreciate this fact. Disregarding the social component of war can lead to erroneous conclusions and from there to future disastrous wars. How societies and cultures fight each other is as dependent upon the composition of the battling societies as it is upon the weapons that those societies use.

Historically, new ideas have always driven innovation. The mastery of ideas, *not* technology, has always been the single greatest determining factor in the changing face of battle. Technology has been the servant of ideas: ideas create concepts, concepts create intellectual structures, and intellectual structures drive technological change. Above all, the single most important influence upon military change has always been the intellectual mastery of new concepts of war. In other words, it has been the interpretation and reinterpretation of Military Theory that has allowed humankind to gain whatever appreciation it has of war's hidden nature.

War is a human enterprise, a social activity. Consequently, nations shape their military theories through the lenses of their histories and their cultures. The resultant theories are, therefore, specifically adapted to their societies. Too often, nations or societies have borrowed someone else's successful military theories, or pieces of it. This dangerous practice comes with a warning: borrowers must respect the unique historical and cultural components of every country's military theory. Where they ignore them, glaring inconsistencies can result. NATO's adoption of the German concept of *Auftragstaktik*, without understanding all the cultural baggage that accompanied the underlying military theories that created it, is an excellent case study. It should also be a warning to all militaries not to attempt to separate strategy and tactics from the theories – or the societies – that created them.

Some societies, like the US, have shown a near religious faith in the ability of technology to unmask the hidden nature of war. This belief has been a growing trend in all Western societies from the time of Newton. Americans are merely the logical result of centuries of growing reliance upon, faith in, and devotion to technological solutions to what are, in fact, human problems. An over reliance upon a scientific paradigm of war can lead to the mistaken belief that human society can become disengaged from war, that the problems of war are not political or social but technological.

Finally, because of the human and social components of war, there is a constant risk of misinterpreting the lessons of one war to be valid for all subsequent wars. Overlooking war's social components can result in military theories that are irrelevant to any situation other than the specific one that created it. The interpretations of Napoleon's strategies and tactics, without linking them to their underlying theories and social supports, are a perfect example of this process. The conviction that war could be the prime tool of diplomacy and that the single climactic battle could be the most important tool of the great commander was just such a disconnected interpretation. It was this misconception that led to a disastrous century long quest for decisive 'Napoleonic' victories from 1815 to 1918.

Notes

1. Roman Jarymowycz, "The Quest for Operational Maneuver in the Normandy Campaign: Simonds and Montgomery Attempt the Armoured Breakout", PhD diss., (McGill University, 1997), 27-28.
2. Williamson Murray, "Armored Warfare: The British, French and German Experiences", Military Innovation in the Interwar Period, ed. by Williamson Murray and Allan R. Millet, (New York, NY, 1996), 45.
3. Douglas A. MacGregor, *Breaking the Phalanx*, (Westport, CN: Praeger, 1997), 40.
4. Michael Howard, "Strategy and Policy in Twentieth – Century Warfare", Harmon Memorial Lecture No. 9, (Colorado Springs, CO: United States Air Force Academy, 5 May 1967), 3.
5. Walter Görlitz, *History of the German General Staff 1657-1945*, Brian Battershaw translator, (New York, NY: Praeger, 1953), 19.
6. Paragraph 302, *Tactics: The Armoured Division and Brigade in Battle*, British Army Command and Staff College, (Camberley, 1986).
7. Ibid.
8. Robert N. Ellithorpe, "Warfare in Transition? American Military Culture Prepares for the Information Age," a presentation for the Biennial International Conference of the Inter-University Seminar on Armed Forces and Society, (Baltimore, MD, 24-26 October 1997), 18.
9. Ryan Henry and C. Edward Peartree, "Military Theory and Information Warfare", Parameters, (Autumn, 1998), 130.
10. Robert P. Pellegrini, *The Links between Science, Philosophy, and Military Theory, Monograph, School of Advanced Airpower Studies*, (Maxwell Air Force Base, AB, 1997), 9.
11. Azar Gat, *The Origins of Military Thought from the Enlightenment to Clausewitz* (Oxford: Oxford University Press, 1989), 29.
12. Michael Howard, *Clausewitz* (Oxford: Oxford University Press, 1983), 13.
13. Howard, Harmon Memorial Lecture 9, 3.
14. Ibid.

> "Do make it clear that generalship, at least in my case, came not by instinct, unsought, but by understanding, hard study and brain-concentration... For my strategy, I could find no teachers in the field: behind me there were some years of military reading... With 2,000 years of examples behind us we have no excuse when fighting, for not fighting well.

T. E. Lawrence
in a letter to BH Liddell Hart

WAR ON LAND

Introduction

HAVING INVESTIGATED A NEW SET OF PARAMETERS with which to construct a model to help us understand the hidden nature of war, how theory affects all aspects of human interaction, and how our understanding and conduct of war is a social phenomenon, we now turn our attention the most important and influential military philosophers, theorists, and strategists. We begin with war on land since it is the oldest form of war. Subsequently we will investigate war at sea and war in the air. The intent of this chapter is to demonstrate two things: First, that there is a necessary difference among the three types of military thinkers: philosophers, theorists, and strategists, that they are distinct and not interchangeable. The theorist builds his ideas upon those of the philosopher; and the strategist develops his thoughts based upon those of the theorist. Second, even though Military Theory is an intricate and growing matrix of thoughts, beliefs, and practices, most of the men discussed repeatedly plow the same intellectual ground. Rarely is an idea or action truly new. More often the idea is an improvement of an earlier premise, the perfection of a previous action, a better explanation of an old idea.

Clearly, the quantity of material regarding war on land dwarfs that available regarding war in the air and at sea and we cannot review all the material. Volume dictates the need to make choices. Subsequently, the list of men discussed in this chapter is a personal one and I accept responsibility for not discussing certain individuals. I also appreciate that his selection leaves gaps. Nonetheless, all cannot be discussed. If we look at Mao Zedong (1893-1976), or Ho Chi Minh (1890-1969) we can legitimately argue that Mao's concept of the soldier being the 'small fish' that feeds in the ocean of society (*Guerrilla Warfare*) is an extension of Sun Tzu, as are Ho Chi Minh's thoughts, and that their contribution to thought on warfare is arguably limited. Similarly, T.E. Lawrence (1888-1935) was an innovator and iconoclast but his book *The Seven Pillars of Wisdom*, however interesting and innovative regarding guerrilla and irregular warfare, is culturally based and experiential and had limited impact on any military theories. Mikhail Tukhachevsky (1893-1937), a founding member of the Red Army, was certainly a visionary who favoured the creation of tank armies and the development of aviation, but did he bring anything to the table that others did not? German general Colmar von der Goltz (1843-1916) must suffer the same fate although both were influential in their respective armies. The list goes on.

This chapter is divided in three thematic parts. Some military thinkers were philosophers interested in building the framework within which military action occurred. Their ideas, like the drawings of Leonardo da Vinci, were bounded neither by known technology nor by reality. They dealt with their subject not as it was but as it *should* be. Some military thinkers were theorists, attempting to create an abstract construct to explain what they had observed. Like theorists who have struggled to explain the physical universe, they assembled premises to aid in understanding. Lastly, there were the practitioners or strategists. Their quest was to understand war sufficiently to be able to perfect and apply their knowledge to gain victory in actual combat. The chapter looks at each group in turn. Individuals are not investigated as they arise but rather chronologically within each group: First the philosophers will be discussed in chronological order; second, starting once more at the beginning of the timeline, the theorists will be discussed; last, returning once again to the beginning, select practitioners or strategists will be investigated. By comparing individuals within their group, a better appreciation may be gained of the distinction among philosophers, theorists, and strategists.

Commanders cannot guarantee victory. Despite their convictions that the gods intervened in all human affairs, the ancients trained incessantly for war. The cost of defeat was too high. The same is true today; losing a war can result in subjugation, loss of sovereignty, even the complete restructuring of the loser's society. Consequently, what the intellectual history of warfare has represented is the quest to uncover the secret of achieving victory. This pursuit has been ongoing for all remembered history. Yet, despite any periodic and recurrent claims to the contrary, no certain solution has ever been found. Battles have always been confused cacophonies of death and destruction, for there are no such things as bloodless battles. For the individual soldier, caught in the maelstrom of man, machine and fire, sheer survival may be goal enough. However, for the leader, victory alone can be the goal; no commander ever goes to battle hoping for a draw. The resources required, combined with the potential consequences of failure keep the stakes high, and although the weapons, geography and reasons for combat may change, the essentials of war remain unchanged and the search for the secret of victory continues.

Our complex matrix ties together the past, present, and future. From ancient times to the present, soldiers have always used the past as a basis upon which to build:

This style of warfare can be traced back to antiquity. Indeed, from the fighting of early Greece to the industrial wars of the 20th century, there is a certain continuity of Western military practice. Greek phalangites and American mounted infantry are linked by Hellenic characteristics of battle: superior discipline, matchless weapons, egalitarian camaraderie, individual initiative, tactical flexibility and a preference for shock battle. The Spartan general, Brasidas, once dismissed the tribes of Illyria because they could not endure shock battle, and this belief in face-to-face battle is found in the Western way of warfare today.[1]

As well as informing the present, the past likewise guides the future. By appreciating how our matrix has been woven, we may better understand how to establish our intellectual foundation.

Like all societies, the ancient Greeks, Macedonians, Egyptians, and Romans all had military theories, but they were not really theories *per se*. The ancients understood war as a natural condition of humankind. It was as common a behaviour as any other social pursuit. The construction of a unifying theory or the collection of principles was unimportant compared with the practical knowledge required to fight. Rather than any working military theories, most of what remains of ancient military practice are reconstructions pieced together by archaeologists and sociologists used to explain social behaviour. The historical record is little more than collected texts on tactics, administration, structure, and training – how to throw up an embankment, how to form a defensive line or how not to let an enemy choose the time and place of battle – rather than a structured theory.

Of all the ancients, arguably, the Romans left the greatest legacy regarding military thought and theory. Even so, what is understood today of Roman warfare is a practical guide. Romans were the first to legislate written laws of war in accordance with their religious and social mores and they observed an extremely formalistic and juridical spirit. They had a deep respect for words [*religio verborum*]. But the Romans did not leave their mark directly. The weight of their forms of war has been felt primarily through the Roman influence upon Medieval and Renaissance writers. Over time, there occasionally emerged new ideas among military writers, but most often, the views expressed, although claiming to be new, were a rehash of previous, often ancient, texts:

> Original ideas appeared in some theoretical literature, but most theories either borrowed from other fields or from other nations... The military schools served as a focal point of development, for a prime opportunity for thought and reflection leading to a better understanding of war existed within the service schools. Furthermore, the need for comprehensive texts and other written references stimulated some instructors to write, and many

others published their lectures either after they retired from teaching or after their views became well accepted. Books on the theory of war proliferated like the schools that spawned them, but the books sometimes generated more criticism than learning.[2]

Some of the individuals we will look at in this chapter could be included in any or all the groups. Where an individual could be considered in more than one cohort, he has been placed where he made the greatest impact. It is not possible to do justice to all the ideas of the men discussed so the focus must be upon their influence, their contribution, or their attempts to raise the level of understanding of the nature of war while keeping in mind that its forms, or character, constantly evolve. Accepting that a representative sampling must suffice, thumbnail sketches of key individuals (or groups) and their impacts are provided, after which there is a general discussion on the interconnectivity of individuals, relationships, philosophies, theories, and strategies.

Philosophers

Sun Tzu (Fourth Century BC)

The first recorded and most enduring philosopher of war remains Sun Tzu. In his treatise *The Art of War,* he guides the reader towards the central premise that understanding war is essential to the well-being of any society: "War is a matter of vital importance to the State; the province of life or death; the road to survival or ruin. It is mandatory that it be thoroughly studied."[3] Little is known about Sun Tzu except when he lived and that he was a general. Even his name is clouded with doubt. His writings were introduced to Europe shortly before the French Revolution by a French Jesuit, Father J. J. M. Amiot. *The Art of War* includes many commentaries by later Chinese philosophers, and it is highly likely that one person wrote the core text, and that others added to it. The exact dates are uncertain, though most scholars now support a time early in the Warring States period (c.453-221 B.C.). Irrespective of who the actual author was, the work has deeply influenced all Eastern military thinking and has enjoyed growing popularity in the West in the last half century. *The Art of War* stresses the unpredictability of battle, the importance of deception and surprise, the close relationship between politics and military policy, and the high social costs of war.

The futility of seeking hard and fast rules and the subtle paradoxes of success are major themes. For instance, the greatest victory, Sun Tzu says, is the battle that is won without being fought. He does not stress the use of specific weaponry or of any technology. Rather, he stresses the philosophical, intellectual, and non-physical aspects of war's nature. Clearly Sun Tzu supports the concept of war being an art, as his title suggests. He practices poetic or artistic imagery to describe its conduct, using an artistic analogy reminding the reader that with only five primary colours, the resultant combinations are infinite and impossible to visualize. Like many ancient teachers attempting to enlighten their pupils, Sun Tzu makes heavy use of metaphor, for instance expressing that war, like water, seeks its own shape with changing conditions. Or, telling the student that the warrior moves like wind or lightning, depending upon the need. Here we can observe the beginning of a trend and perhaps a thread that ties together the foundations of all subsequent theories. The use of metaphor builds an internal tension in the mind of the student. War is always described as something to which the pupil can relate innately, but which may not have any direct correlation to war itself, like the comparison of war to colour and to water. This tension can be discerned in all subsequent theoretical works on war as each new author attempts to develop a theory based on empirical, historical evidence.[4]

In summary, Sun Tzu's *The Art of War*, whether written by him or by some combination of writers known to us by his name, created the foundation upon which much of military philosophy rests. Although unknown in the West for one and a half millennia, once discovered, it has never been put aside. The philosophical nature of the work, with its paradoxes and internal tensions, has given it a timelessness not shared by any other text. Even in the twenty-first century, *The Art of War* remains rightly the cornerstone of every professional military library.

Niccolò Machiavelli (1469-1527)

The most famous and influential philosopher to seek to re-establish the Roman way of war was this Florentine civil servant:

> Machiavelli's writings marked the beginning of nearly three centuries of military thought that was strongly influenced by classical thought, that reflected an interest in the search for principle, fundamentals, general rules, or any of the variety synonyms used to define such basic concepts, and that increasingly was influenced by technological growth and scientific inquiry.[5]

He was the author of several treatises on war, warfare, and politics: although most celebrated for *Il Principe* (*The Prince*), he also wrote *Arte della Guerra* (*Art of War*) – which he mostly lifted from Vegetius – and *Discorsi* (*Discourses on Titus Livy*). His was a political outlook and he is, therefore, considered as the father of modern political science. Although, he studied war as it related to soldiers and fighting, for Machiavelli, war existed in the political realm. Thus, that war was just, which was necessary.[6] A prince, said Machiavelli, needed courage, determination, and fortitude, a list harkening back to Roman *virtù*. The state was the most important facet of society, and in the defence of his state, a prince could violate any law of humanity with justification, a concept that had his writings placed on the Roman Church's *Index* of prohibited texts in 1559. Renowned for his rationalization of almost any atrocity for the sake of protecting the state, it is in his works that we see the first modern arguments that war is a serious business, worthy of study as an integral part of civic duty. His refutation of the use of mercenary *condottieri* is a milestone in establishing the basis for a civil-military relationship that holds true to this day for Machiavelli believed mercenaries to be useless and dangerous. If a prince founded his state on these arms, he would never be either firm or safe. Because these men fought only so long as there was cash on hand, no prince could put his faith, nor the fate of his state, in their cowardly hands, or as a later expression explained, *pas d'argent, pas de Suisse*.

Although highly unlikely that he was familiar with Sun Tzu, Machiavelli, too, continued in the tradition of stressing the intellectual, and the moral over the physical. His strong belief that a prince should publicly demonstrate his religiosity to his subjects, irrespective of his personal beliefs, demonstrated an understanding of the power of moral ascendancy over brute force. In his *Arte della Guerra*, he offered the sound advice to the commander that he should always endeavour to maintain the initiative. He established forever the concept of *raison d'état* and his attempts to systematize his ideas on war and politics made him *primus inter pares* of Western military philosophers.

In summary, Machiavelli remains the point of departure for military philosophy in the West. His insistence upon attempting to understand the nature of war as the underlying basis of all study of warfare has been carried forward to the present day. His writings not only form the foundation of modern political science, but his military philosophy also creates the underpinning of most of the theorists who followed. Although vilified and banned by the church, his books continued to be read and studied by soldiers and statesmen alike from the time of their writing until the present day.

Miyamoto Musashi (1584-1645)

There is another Eastern treatise that deserves mention, even if its full influence has not yet been felt. The Japanese *Book of Five Rings*, like the Chinese *Art of War*, is a philosophical, Zen-based treatise on combat. The tract is formed around the conceptual construct of nature's five elements: fire, water, wind, earth, and void. Like its Chinese predecessor, the importance of technology is almost completely ignored in this Japanese text and the reader is encouraged to seek enlightenment and understanding of combat through self-perfection and in attempting to imitate the qualities of the natural elements. For the average Western student of war, this Zen approach is difficult if not impossible to grasp. Nonetheless, the internal tension and the philosophical foundations are a natural extension of Sun Tzu.

Musashi was a follower of the ancient *Samurai* code of *Bushido* or the 'way of the warrior'. At age fifty-nine, he retired to a cave to contemplate his personal success and wrote *Book of Five Rings*. The work is divided into five parts or books, one for each of the five Zen elements: the first book (Ground) introduces Musashi's definition of strategy; the second (Water) details the fencing techniques of using the sword; the third (Fire) analyses strategies of field combat; the fourth (Wind) compares Musashi's personal thoughts on strategy with others prevalent in Medieval Japan; and the last (Void) is a philosophical and epistemological dissertation attempting to explain knowledge.

Although *Five Rings* became a standard text among Samurai and so entered the Japanese military, it was practically unknown in the West before the twentieth century. Since the first English translation in 1974, the work has become an underground classic in the American business community and modern military theorists like John Boyd, John Warden, Bill Lind, and the proponents of 4th Generation Warfare, have all attempted to embrace it. Musashi's influence has been felt at the higher levels of command in both the American and the Canadian militaries as *Five Rings* continues to be read and discussed at staff colleges and war colleges.

It is difficult to gauge the military impact of *Five Rings* because it has been studied in the West for only a few decades. Certainly, it has had a profound influence in Japan. Although not in the same category as Sun Tzu or Machiavelli, Musashi deserves mention because of his amplification of Sun Tzu's *Taoist* philosophy, which can be discerned in much of the modern ideas of theorists like JFC Fuller, Liddell Hart, Richard Simpkin, Boyd, and Warden, whether acknowledged or not.

Theorists

Flavius Vegetius Renatus (Fourth Century AD)

Until Clausewitz's *On War* appeared in 1832, to guide those who would understand the nature of Napoleonic warfare, no single writer in the West was more influential than the Roman historian and writer Flavius Vegetius Renatus. The language of warfare in the modern era has its roots in his five-volume *De Re Militari*, also known as *Epitoma rei militaris*, which reputedly influenced commanders from Richard the Lionheart and Charlemagne to Raimondo Montecuccoli. Vegetius' treatise was written late in the fourth century, in the hopes of revitalizing the Roman army after its disastrous defeat at the Battle of Adrianople in 378. Although little is known about him personally, he is frequently mentioned in the works of the Great Captains of war and his books were the last word in all military matters:

> The most influential treatise on war to survive from the era of Roman domination is Vegetius's *De Re Militari* (*On Military Institutions*). Written during the decline of the Western empire but extolling the virtues of early Rome, it was used little by the Romans. During the Renaissance, however, nearly a millennium later, *De Re Militari* was the most popular work, and

possibly the only work widely used, on the practice of war. Vegetius, like every other serious writer on military topics, well understood the importance of victory in war and attempted to point out the fundamentals essential to it. Vegetius included neither enumerations nor terse lists, however, focusing instead on discipline, organization, training, and administration.[7]

De Re Militari was written not about the army that Vegetius knew but rather of an idealized version of previous, more successful Roman legions. Among his are men who predate Rome's fall, men such as Cato, Augustus, Trajan, and Hadrian. Vegetius' work was a 'handbook on war' and studied diligently by the warrior-kings of the Middle Ages. It stressed hard work, iron discipline and thorough training. The great readability of the work, the fact that it was filled with pithy 'rules', along with the direct link of this work with the glory of Imperial Rome gave the text a lasting popularity. Rulers valued the work greatly "but they particularly valued the 26 chapters on strategy, tactics, and the principles of war (or military procedure) contained in Book III."[8] Machiavelli became interested in *De Re Militari* and used it as the basis for his *Arte della Guerra*, copying large portions of the text virtually verbatim.

In summary, Vegetius re-introduced to the West the military arts and sciences of the Roman way of war and in the process became *the* source for soldiers to study for over a thousand years. In modern terms, his work is almost a cookbook of rules and regulations. Nonetheless, *De Re Militari* was the authoritative text used by most Europe's warrior-kings and became the point of departure for the writings of Machiavelli, Montecuccoli, and Marshall de Saxe. Through their influences, his ideas were passed to countless others.

Raimondo Montecuccoli (1608-1681)

This Italian-born Austrian general and soldier-theorist was the chief founder of the Austrian army. *Il duca* di Melfi and *generalissimo* was an extremely successful soldier and one of the most popular writers about war in his lifetime. His greatest work was *Trattato della guerra*, written 1639-1643. He believed that a set of principles could be uncovered to conduct war in a more practical manner and believed war to be a science and so stressed mathematics and physics, which had an impact upon his engineer readers like Henry Lloyd.

Montecuccoli applied the Aristotelian method to war. By observation and experience, he attempted to formulate a system for the understanding of war. "He intended to reveal war's near-mystical mechanics, symbolized by universal principles, rules, and maxims that anyone could use when conducting military campaigns."[9] Montecuccoli did not subscribe to Machiavelli's disdain for professional armies. He was a firm advocate of standing professional armies of the type pioneered by Maurice of Nassau. He developed a military organization based upon a strong combined arms standing army, equipped, and trained for war. He drew his inspiration from the belief that the state was the only legitimate authority to declare and prosecute war, greatly influenced by the political philosophy of the Flemish Justus Lipsius' *Constantia* as well as the concept of *virtù* from Machiavelli's *Discorsi*. He expounded upon tactics for both infantry and cavalry, emphasizing the need to seize and maintain the initiative and articulated a strategy for limited operations as well as wars of attrition. He distinguished between external war (between states) and internal war (civil war).

In summary, Montecuccoli made his mark as a commander, an administrator and as a theorist. His thoughts on the civil-military relationship owe much to Machiavelli and pre-date Clausewitz's famous dictum connecting war and politics. Although he, like Henry Lloyd and Jomini after him, attempted to formulate scientific principles for the conduct of war, he eventually conceded that although containing elements of science, war was an art.[10] Widely read and highly respected, he was a major influence during the latter half of the seventeenth century.

Henry Lloyd (1729-1783)

Lloyd was Welsh by birth but became European, leaving his native land at an early age in search of adventure. He served in the French, Prussian, Brunswickian, Austrian and Russian armies and even commanded a Russian division. Lloyd was a soldier who became an intellectual, a *philosophe*, and though his influence was widespread in his own era, he has now been relegated to the footnotes of history. Born to a landed gentleman, Lloyd's father sent him to Oxford for his education and thence to the Jesuits, who sent him to Roman College where he was instilled with the mathematics, and physics of Isaac Newton.

Lloyd's non-noble birth, combined with his technical education, positioned him well to become a military engineer. He joined the French army, fighting under de Saxe in the *Corps des Ingénieurs de Génie Militaire*. Not surprisingly, Lloyd later borrowed heavily from de Saxe's *Mes rêveries; ou Mémoires sur l'art de la guerre* to formulate his own thoughts on equipment, drill, discipline, and tactics: "[O]f all the mechanical parts of war, none is more essential than that of *marching*."[11] His time under de Saxe's command was formative of his military theories:

> It is universally agreed upon, that no art or science is more difficult, than that of war; yet by an unaccountable contradiction of the human mind, those who embrace this profession take little or no pains to study it. They seem to think, that the knowledge of a few insignificant and useless trifles constitutes a great officer. This opinion is so general, that little or nothing is taught at present in any army whatsoever. The continual changes and variety of motions, evolutions, etc., which the soldiers are taught, prove evidently, they are founded on mere caprice. This art, like all others, is founded on certain fixed

principles, which are by their nature invariable; the application of them can only be varied: but they are themselves constant."[12]

The engineer's perspective can be discerned in Lloyd's writings as he puts great emphasis of physical terrain: "Not only an exact knowledge must be had of all fortified towns, but even of all the villages...because they form defiles, which being occupied put an effectual stop to the enemy."[13] Lloyd sought to understand the nature of war, being the first European to combine, as he put it, two "classes; Didactical [experiential] and Historical [theoretical]: the first are of great use, no doubt, but by no means comparable to the others."[14] In his writing, Lloyd established a new paradigm for the study of war, which he expounded upon in the Preface to *Continuation of the History of the Late War in Germany between the King of Prussia and the Empress of Germany and her Allies*. He stated that the art of war was founded upon certain fixed and constant principles; explained the military operations (including maps); gave a military description of the 'feat of war;' and gave a description of geography and terrain.

In summary, although now nearly forgotten, Lloyd was both widely read and greatly respected until he was overshadowed by the likes of Clausewitz and Jomini, his influence being much stronger on the latter than on the former. Napoleon read his works and Jomini declared that his greatest intellectual debt was owed to Lloyd. Jomini's insistence that war was dependent upon certain immutable principles was clearly an idea taken from Lloyd's work.

Scharnhorst, Gneisenau and the Prussian Reformers

The calamitous defeat of the Prussians at the Battle of Jena in 1806, forced the Prussia to look deeply into itself and to change. To appreciate the changes after Jena, some background is useful. The two most distinguished and influential military critics in Prussia before Jena were Georg Heinrich von Behrenhorst (1733-1814) and his pupil Heinrich von Bülow (1757-1807). Behrenhorst published *Betrachtung* über *die Kriegskunst* (*Dissertation on the Art of War*) in 1797 that "heralded the revolution in military thinking which was to match the revolution in warfare."[15] Unlike Archduke Charles and Lloyd, he did not believe in principles. Bülow published *Geist des neueren Kriegssystems* (*Heart of the New System of War*) in 1799; he was trying to bring order to the chaos of war and an attempt to codify contemporary practices. Both Behrenhorst and Bülow were widely read in Prussia but they tended to ignore "much of the innovation of the armies of Revolutionary France."[16] Both called for more freedom for individual soldiers and argued for the acceptance of the importance of the moral, and spiritual, forces in war. They had a "profound influence on Clausewitz, and even the famous aphorism that war is merely the continuation of politics by other means is based on one of Bülow's dicta."[17] The early influence of this intellectual intercourse upon Clausewitz cannot be overstated.

WAR ON LAND

The wars of the French Revolution wrought radical changes to European warfare and by the 1790s, there were calls across Europe for reform. Many officers wondered whether the new tactics, techniques and structures being introduced by France should not be adopted in Germany. Among the various officers' societies, especially Berlin's *Militärische Gesellschaft*, the debate was active and at times heated. Among the most active of those calling for change was a Hanoverian officer, Scharnhorst. The son of a tenant farmer, he had been a student at the academy of Friedrich Wilhelm Ernst *Graf* zu Schaumburg-Lippe-Bückeburg. The *Graf* was a true son of the *Aufklärung* (German Enlightenment) and taught all his pupils that the mastery of the art of war required much more than Frederician parade-drill and blind obedience. He taught his young charges to broaden their minds in terms of culture, linguistics, and the humanities as well as the accepted military subjects.[18] Scharnhorst became a prolific writer, publishing handbooks, field manuals and military studies on the reasons for the success of the French armies, gaining a well-deserved reputation for being a leading military theorist. But his reputation also extended to combat: In 1793 during War of the First Coalition; in an engagement against the French, he took command of several infantry units intent on fleeing, halted them, then coordinated an effective rearguard action, conducting an orderly withdrawal that helped preserve his entire corps.[19] His leadership, all without orders, caught the attention of the Prussian king.

In 1801, he accepted a commission in the Prussian army and was posted to the *Generalquartiermeisterstab* in Berlin where he joined the *Militärische Gesellschaft* where he immediately became its head. One of Scharnhorst's duties on the Berlin staff was to oversee military education. This made him the director of Berlin's *Militärakademie*, with the Prussian crown prince as one his students. At the helm of the *Gesellschaft*, Scharnhorst guided the active discussions suggesting exactly what type of reforms were needed in Prussia. Some called for increased freedom of action at lower ranks, as well as the importance of the moral aspects of war – both ideas being fundamental to the future development of *Auftragstaktik*.

After Jena, the king struck a commission, headed by Scharnhorst, to rebuild and reform the Prussian army. The reformers, fresh from their studies of the French Revolution and still stinging from Napoleon's humiliation of their army at Jena, set about to rejuvenate their army. The theoretical writings of Behrenhorst and Bülow became the intellectual underpinnings of the Reform Commission's attempts to create a new system of warfare. The ideas found there, as well as in the *Militärische Gesellschaft*, allowed Scharnhorst to lay the foundations of a new theory of warfare, one based upon the individual worth of commanders at all levels and independent action aided by a highly trained and proficient general staff. The commission's work would create a self-sustaining theoretical model and set in train a series of reforms that, arguably, continue – through the legacy of Clausewitz – to the present-day Bundeswehr and beyond.

STRATEGIA

Baron Antoine-Henri de Jomini (1779-1869)

In contrast to Clausewitz, Jomini's thoughts on war became the touchstone of those who espoused a finite set of principles of war. French military thought flourished under the guiding hand of the Jominian outlook while across the border, the German schools did the same under the Clausewitzian model.

The writings of Jomini and their use of science as a basis for understanding war reached its theoretical apogee during the Industrial Revolution. This erstwhile Swiss banker and amateur soldier eventually rose to serve as Chief of Staff to Marshal Michel Ney, Napoleon's *Chef de Cavalerie*. Having read Lloyd, Jomini analyzed Napoleon's Italian campaigns in a search for unchanging principles. Upon reading Jomini's first treatise, *Précis de l'art de la guerre*, Napoleon is said to have exclaimed that Jomini had discovered what was in his mind. Jomini believed that his mathematical formulae laid out the proper organization of military formations, as well as the size and the direction of an attack at the 'decisive point.'

A fundamental challenge facing every military theorist is to reduce the assorted forces with which a theory must contend to a manageable level. For Jomini, the solution was to reduce war to material battlefield considerations, which were all tangible and to mathematical concepts, which in the were universally true. In this way, he made he believed that he made his concepts universally applicable across the full spectrum of military conflict, much as anything in nature could be reduced to its basic building blocks and thereby studied and understood.[20]

Napoleon's successes produced manifold changes in how war was both fought and studied, and two important trends emerged. The most popular interpreter of Napoleon being Jomini, the belief in the fact that the mastery of a few underlying principles would lead to victory was the immediate consequence. More careful reflection brought the second trend; teachers in military schools began to write books on theory and the influence of the memoirs of successful field commanders began to wane. Both trends were clearly Jominian, and as the American Civil War loomed, it was Jomini who stood at the forefront of all military thinking not only in Europe but also in the US.

In the final analysis, however, Jomini's great popularity was not as enduring as his contemporary rival Clausewitz. Extremely influential, even into the present day, Jomini's overly practical approach to warfare has suffered over time. His 'cookbook of war', contrary to his claims, has become dated:

> Jomini has been accused from time to time of merely offering a cookbook of war, this by persons who do not reflect on how extraordinarily useful a cookbook may be. His basic ideas, many of them influenced by his reading of Henry Lloyd, include interior and exterior lines, the decisive point, concentration of strength against weakness, annihilation of the enemy force,

the primary importance of the offensive, surprise, and the potentially decisive role of logistics. The essential object of all this was to win a favorable result through the concentration of strength against weakness. Jomini felt these were fundamental, almost mathematical principles of war and that they were good for all time.[21]

In summary, Jomini's star, which burned so brightly during his own lifetime and even up to the end of the twentieth century, has grown dim. Clausewitz has taken his place of prominence in the pantheon of military theorists. Perhaps unfairly, modern students of war see Jomini as too imbued with a technical, geometrical, almost mechanical point of view. Even in the US armed forces, where technology has a place of honour as perhaps nowhere else, his influence is now diminished. The concepts of 'lines of operation' and 'principles of war' remain strong in the theoretical lexicon, but they have been subsumed by the Clausewitzian model of particularism and the concept of dichotomy found both in Sun Tzu and in Clausewitz.

Carl Maria von Clausewitz (1780-1831)

In the Western world, no one else has come to be so closely associated with military thought and theory as Clausewitz. Unfortunately, his pervasive influence has acted to blind many professional soldiers and scholars to all other military theorists. Once the US army rediscovered *On War* in the 1980s it was not long before articles, books, dissertations and essays on this author and his work filled the pages of American PME. By itself, this focus would not have been a bad thing, since theories of war had not been a topic of discussion for many years amongst professional officers in the English-speaking world. However, the difficulty was that Clausewitz's ideas were spoken of as revealed truth and not as another theory to add to the pantheon of military thought. Clausewitz was soon discussed in military circles as *the* theorist rather than *one* among many.

As for his ideas, Clausewitz was convinced that war was not a science and that it could not be codified: "Nor can the theory of war apply to the concept of law to action, since no prescriptive formulation universal enough to deserve the name of law can be applied to the constant change and diversity of the phenomena of war." (II: 4) This thinking was in direct opposition Jomini's teachings that war contained a hidden code to be broken, that war could be reduced to a set of principles.

Head and shoulders above his contemporaries, he was fortunate to have been mentored by some of the very best military minds of his age. His inclusion among the reformers, as adjutant to Scharnhorst, would mark him for life. Later, this impact would manifest itself in Clausewitz's own teachings, first during his time as director of the *Allgemeine Kriegsschule* in Berlin, and then, later, when his seminal work, *On War*, became the touchstone not only of *Auftragstaktik*, but also of the German way of war.

STRATEGIA

Clausewitz believed that the Napoleonic Wars had freed warfare from the confined view of the professional soldier, that it had been turned back over to the people. Clausewitz was by no means alone; however, he was the best:

> The military leaders and theorists who reached maturity in the Napoleonic Era developed a comprehensive understanding of – and thus control over – the new forms of war. This theoretical achievement capped all other changes that had occurred in equipment, organization, tactics and strategy. Their recognition of the nature of modern conflict was best expressed in Clausewitz's *On War*.[22]

Nearly every modern military writer cites Clausewitz, at least once, as a matter of routine, in recognition that his influence is so widespread that he must be at least acknowledged to lend credibility to any scholarly discussion regarding war. Perhaps because of this practice, Clausewitz has gained the descriptor 'more often quoted than read.' In many ways, he has become 'an author for all seasons' allowing all and sundry to quote from his tome in absolute assuredness that they are being supported by some part of his unedited writings. It would not be far-fetched to state: "no other author has ever been remotely as influential, and indeed to this day his work forms the cornerstone of modern strategic thought."[23]

Clausewitz's view of war is imbued with not only the writing of Immanuel Kant, but also of the scientific ideas of Galileo and Newton. "Specifically, Kant's concept of space, time, order, and morality, his concept of absolute versus the 'real' world, his system of knowledge, and his concept of genius are all evident in *On War*."[24] Clausewitz's use of the concepts of *absolute war* (as opposed to *real war* or *actual war*) was strongly influenced by the authority of both science and philosophy during the *Aufklärung*. Kant talks about the ideal versus the actual as does Galileo. Galileo had created the 'ideal' state of a vacuum to help him describe the acceleration of falling bodies. When Clausewitz says that "In its absolute form, war, where everything follows from necessary causes and all actions rapidly affect one another, there is, if I may say, no intervening neutral void ..." (VIII, 3) his phrasing is reminiscent of Galileo.

In Book VIII: 2, Clausewitz proposed the absolute form of war as a referential model to describe real war. Since real war was too complex to describe war's true nature, a tool was needed to assist in its understanding:

> It follows that war is dependent on the interplay of possibilities and probabilities, of good and bad luck, conditions in which strictly logical reasoning often plays no part at all and is always apt to be a most unsuitable and awkward intellectual tool. It follows, too, that war can be a matter of degree. Theory must concede all this; but it has the duty to give priority to the absolute form of war and to make that form a general point of reference,

so that he who wants to learn from theory becomes accustomed to keeping that point in view constantly, to measuring all his hopes and fears by it, and to approximating it *when he can* or *when he must*.

Clausewitz objected to the dissection of war it into its component parts, his objection was based on the belief that war, like so much of human activity, was greater than the sum of its elemental parts. He believed that while analysis could be useful, war needed to be understood in context. He insisted that isolated causes and effects could not be clearly established in isolation of the greater whole.

In some ways, Clausewitz's work was a precursor to what is now known as Complex Systems Theory. In *On War*, he included a cogent analysis of the role of theory, particularly as it related to the study of war and his intellectual structure provided a conceptual framework within which to situate military concepts like *center of gravity*, *mass*, and *friction* – terms obviously borrowed from Newtonian physics – *decisive points*, *lines of operations*, *bases of operations* and *culmination*. In Book II he connected this analysis to three closely related issues: whether war most closely approximates an art or a science; the role and characteristics of critical analysis of war; and the use of historical examples to assist in the study of war. He discussed the purpose of theoretical knowledge and its relationship to the conduct of war as well as the relationship between theory and critical analysis as well as the value of historical study.

Over time, Clausewitz came to appreciate that his model of absolute war needed refinement and that limited war was much more common than total war. Unfortunately, Clausewitz died before being able to edit and refine his writings. His work was published posthumously by his widow, which would explain some of the inconsistencies in *On War*. Nonetheless, for good or ill, understood or not, Clausewitz is today 'first among equals' among military theorists in the twenty-first century.

Dennis Hart Mahan (1802-1871) and Henry Wager Halleck (1815-1872)

Unless you are a graduate of the US Military Academy, or a student of American military history, Dennis Hart Mahan is likely an obscure figure. His son, Admiral Alfred Mahan, has long overshadowed him. Graduating at the top of his of West Point class in 1824, the elder Mahan was commissioned into the Army Corps of Engineers and soon made an assistant professor of mathematics at the academy. He travelled to France where he studied from 1826 to 1830 at the French Army School of Engineering and Artillery. Returning to West Point in 1830, he became a professor of engineering, where he developed a course entitled 'Engineering and the Science of War.' He produced a series of texts for his cadets, most of them based on European books. Deeply influenced by Jomini, Mahan taught at West Point continuously until his death in 1871, spreading his thoughts to generations of American army officers as well as to his own son. Despite the lack of original thought in his works, when we consider the years of his tenure and the fact that all major commanders on both sides of the US Civil War had been his students, Mahan's influence cannot be overstated and can be clearly detected in the bloody battles of the US Civil War.

Keeping intellectual company with Mahan was Henry W. Halleck. A favoured student of Mahan and a graduate of the Class of 1839, he was commissioned, like Mahan, into the Corps of Engineers. He was a widely read theoretician, having studied Jomini as well as the Archduke Charles' *Principes de la Stratégie*. He published *Elements of Military Art and Science* in 1846, after returning from a military tour of European fortifications, and translated Jomini's *Life of Napoleon* during the Civil War. His *Elements of Military Art and Science* preceded the written works of Mahan and became a standard text at his alma mater, thereby influencing several generations of American military officers.

Halleck claimed no originality in his work and not surprisingly, focused upon fortifications. His writing had an abstract quality and stressed the various aspects of engineering, conceding that war was both an art and a science but giving precedence to the former:

> War in its most extensive sense may be regarded both as a *science* and an art ... So is engineering a science so far as it investigates the general principles of fortification, and also artillery, in analyzing the principles of gunnery; but both are arts when considered with reference to the practical rules for the construction, attack, and defence of forts, or for the use of cannon.[25]

Carrying on the European tradition from Lloyd, through Jomini, Halleck makes much of *lines*. He describes *lines of defence*, *lines of operations*, *double lines*, *multiple lines*, *interior lines*, *exterior lines*, *concentric lines*, *eccentric lines*, *primary lines*, *secondary lines*, and even *accidental lines*![26] His thoughts on warfare have a highly geometrical aspect and, like Jomini, he used the language of geometry to describe battle. "If the army A is obliged to cover the point *a*, the army B will cover all the space without the circle whose radius is *a*B; and of course A continues to cover the point *a* so long as it remains within the circle *a*B."[27] Although, to be fair, there is more of Archduke Charles than Jomini in his overall thinking, especially where it comes to protecting and attacking *strategic points*.

Halleck was a good engineer and a good administrator, but he was a poor commander. Both contemporaries and later historians criticized him as mediocre in the field, having been replaced twice by General U.S. Grant. Nonetheless, he is important in the literature of American military thought. Like his mentor, Dennis Mahan, he is not important for his original thought but rather for his introduction to the US army of the theories of Europeans like Lloyd, Archduke Charles, and Jomini. Despite the emphasis that both men put upon engineering and fortifications, Russell

WAR ON LAND

Weigley credits both the senior Mahan and Halleck for creating the beginnings of professional military literature in America, especially when we consider that the US army's primary tasks where almost exclusively constabulary.[28] In sum, their influence, although widespread, was limited to the way in which they introduced and imbued the US army with the Jominian School of warfare.

J.F.C. Fuller (1878-1966) and Basil Liddell Hart (1895-1970)

Easily the most important British theorists to emerge after the First World War were Captain Basil Liddell Hart and General J.F.C. 'Boney' Fuller. Both men became early and strong proponents of mechanization and although Liddell Hart was more widely read (he was military correspondent for the London *Times* from 1925 to 1935), Fuller's reputation has weathered better than his junior contemporary's whose views have lost much of their favour over the years. From a theoretical perspective, both men envisaged what is now popularly referred to as strategic paralysis. Their greatest influence was outside of their native country, being mostly in Germany, and has been cited by at least one German *Panzer* general as being fundamental to his development of armoured warfare.

Liddell Hart is most famous for his theory of the 'indirect approach'. It was an attempt to avoid frontal warfare, a tactic with which he was familiar from the First World War. In essence, the idea was to avoid strength and attack a weaker part of the enemy by going around him. Although trumpeted by him as original thinking, it was far from it. This concept can be found in both Sun Tzu and Musashi. It can also be found in the German 'von Hutier tactics' named after General Oskar von Hutier, (1857-1934) Commander of the German 8th Army during the First World War. Liddell Hart was steadfastly critical of Clausewitz or the 'Mahdi of Mass' as he called him, seeing him as a proponent of bloody battle and mass armies clashing with each other unnecessarily, although there is now some doubt whether Liddell Hart ever really studied *On War*. Liddell Hart's thinking was centred on the concept of sparing Great Britain from another bloodletting like the First World War. His theory, therefore, once translated into strategy, can be described as a strategy of paralysis, and despite his disdain for the Prussian, some of Clausewitz crept into his language, if not his thinking:

> A strategist should think in terms of paralysing, not of killing. Even on the lower plane of warfare, a man killed is merely one man less, whereas a man unnerved is a highly infectious carrier of fear, capable of spreading an epidemic of panic. On a higher plane of warfare, the impression made on the mind of the opposing commander can nullify the whole fighting power his troops possess. And on a still higher plane, psychological pressure on the government of a country may suffice to cancel all the resources at its command – so that

the sword drops from a paralysed hand.[29]

Although contemporaries, Fuller's experience was deeper and broader than Liddell Hart's. He subscribed to the concepts of British philosopher Herbert Spencer; therefore, his theories were founded upon a synthesis of ideas. Further, this Spencerian philosophical background led Fuller to study war not as an art but as a science. Intelligent, literate, and bellicose, 'Boney' was disdainful of many of his brother British officers. He was the conceptual architect of the first large scale tank battle at Cambrai in 1917 and the author of Plan 1919, heralded by some as the precursor to *Blitzkrieg*. Fuller, like Liddell Hart, was a prolific writer and through his articles, books, essays, and commentaries came to be considered the 'brain' of the British tank corps.

Like Liddell Hart, he was critical of Clausewitz. But Fuller seems to have adopted at least some of Clausewitz's thinking into his own. In *The Foundations of the Science of War*, his examination of the nature of war, he introduces the concept of a trilogy. His threefold order was "a foundation so universal that it may be considered axiomatic to knowledge in all its forms."[30] Fuller posited three spheres of war: physical, mental, and moral, a categorization, which was notably like Clausewitz's famed 'wondrous trinity' of armed forces (physical), government (mental), and population (moral). He believed that "The physical strength of an army lies in its organization, controlled by its brain. Paralyse this brain and the body ceases to operate."[31] This belief is clearly reminiscent of Spenser Wilkinson's 1890, *The Brain of an Army*. Fuller insisted that such 'brain warfare' was the most effective and efficient way to destroy the enemy's military organization and hence its military strength. To economize the application of military force, it was far better to produce the instantaneous effects of a 'shot through the head,' rather than the slow bleed of successive, non-lethal body wounds. Shades of Sun Tzu.

The introduction of tanks to the battlefield stimulated theoretical work on how they should be employed. Although England's Fuller and Liddell Hart led the way, it was in their home country that they also encountered the greatest resistance to their ideas. Ironically, it was Heinz Guderian's reading of Fuller and Liddell Hart that stimulated his interests in the use of armour as independent formations on the battlefield and not just in support of infantry. The Soviets also read Fuller and Tukhachevsky was influenced by Fuller's ideas of mechanized and air force manoeuvres, annihilating an enemy by achieving faster mobility than he could sustain.

In summary, the interwar period saw the rise of two British theorists, both of whom were deeply influenced by their experiences during the First World War. Although creating distinct theories – Fuller was tactically focused whereas Liddell Hart was more strategically oriented – their thoughts were related. Both sought to free their state, and warfare generally, from the shackles of a style of warfare based on bleeding an opponent white. Their ideas on mechanization, the indirect approach and strategic paralysis, although less ardently preached, remain valid. Liddell Hart

WAR ON LAND

could justly be accused of exaggerating the value of the indirect approach and Fuller suffered somewhat similar narrowness in his advocacy of tanks being the only decisive manoeuvre force. Nonetheless, together their theories had a direct effect upon the German *Panzer* generals and therefore upon how war was fought from 1939 onwards.

Huba Wass de Czege (1941-) and American Intellectual Rebirth

With the close of the Second World War, America became the world's first superpower. The Cold War caused a hiatus in military thought other than that related to nuclear deterrence, which was primarily related to air power theory. As the twentieth century ended, however, military thought came out of its slumber. A Hungarian-born US infantry officer, West Point and Harvard graduate, helped to lead the way. Huba Wass de Czege was the intellectual father of the American doctrine of AirLand Battle as well as the founder of the US Army School of Advanced Military Studies (SAMS), in which selected staff college graduates study military history, thought and theory. Wass de Czege, as a lieutenant colonel, was the chief writer responsible for the drafting of US Army FM 100-5, *Operations* in 1982. The manual was the keystone document of army doctrine and introduced AirLand Battle, which replaced the previous edition's doctrine of attrition and firepower with rediscovered concepts based on maneuver and deep-strike offensives and closely integrated air power with land power.

The following year, Wass de Czege became the first Director of SAMS in Leavenworth, taking the best graduates from the adjacent staff college course and inculcating them with the new doctrine during the year long course. But Wass de Czege was not just an academic. He had battlefield credibility, having served two combat tours in Vietnam:

> Huba was a wonderful company commander, and he had excellent credentials ... This was his second tour in Viet Nam, so he knew his way around. He was the bravest and smartest man in his company and personally manned an ambush position every night. His company headquarters, which amounted to him, his radio operators, his first sergeant and his forward observer, had killed and wounded more enemy than the entire rest of the battalion.[32]

Like Generals Don Holder and Donn A. Starry, both of whom were principal drivers in supporting concept development for the US army, Wass de Czege was a warrior-scholar, a member of the generation of army officers that had seen the disasters firsthand and were eager to change the way the US army thought about combat.

The new doctrine was readily absorbed by the students and in a methodology reminiscent of the early days of the Prussian General Staff's promulgation of *Auftragstaktik*, the graduates inculcated the entire US army. When General Norman Schwarzkopf led the invasion of Kuwait, most of his planning staff consisted of Wass de Czege acolytes – the so-called 'Jedi Knights.' The influence was obvious in the multiple thrust, manoeuvrist, coordinated air-land, plan.

Wass de Czege is a classicist, and you can quickly determine the influence of Clausewitz in his writing. Observe, for instance, his response to the claim that air power had won the Kosovo campaign:

> War is first and foremost a contest of wills and the enemy quits not because of what has already happened, but because of what he believes might happen if he doesn't. Fires, whether standoff or close, are transient. They have great moral influence, but only for the duration of their existence. Extended range fires can set the terms of close combat, but the enemy quits because he fears the inevitability of defeat. There is no surer way to demonstrate that inevitability than with an overwhelming and imminent threat on the ground. ... Slobodan Milosevic finally caved, not because of the 77 days of precision bombing, but because he became convinced that NATO would ultimately launch a ground campaign in spite of earlier assurances to the contrary.[33]

In summary, Wass de Czege's importance and influence, like Halleck's before him, came not so much in his original thought as in his influence upon the US army. Beginning with the 1982 re-write of the army's keystone doctrine manual and followed closely thereupon by the leadership he gave the army's School of Advanced Military Studies, he put the army on a path towards reinvigorating its institutional appreciation of military thought. All officers, from second lieutenant to general, were made to appreciate that, as military professionals, they had an obligation to study Military Theory. His influence continues to be felt across the generations of army officers, who either attended SAMS or worked for SAMS graduates.

Strategists

There is a general acceptance among military historians that from the mid 16th to the mid 17th century the West underwent what Professor Michael Roberts coined a 'Military Revolution.' Before this time, the greater part of military thought and theory comprised reading of the exploits of Alexander of Macedon (356-323 BC) or studying Julius Caesar's (c101-44 BC) *de Bello Gallico*. These two soldier-statesmen along with some of their brother Great Captains became near-mythological in their greatness and in some cases their social and political influences can still be felt. But from the perspective of military thinking, there was little significant shift in warfare until Maurice of Nassau, who having read Vegetius and likely Machiavelli, changed the way Europeans went to war. Thus, although regarded as epitomes of strategy, inspirational leadership, tactics, law, and administration Alexander, Darius the Great, Caesar, Hannibal and their like will not be discussed here.

Maurice of Nassau (1567-1625) and Gustavus Adolphus (1595-1632)

In the 15th and 16th centuries, many armies stopped fighting each other. Captains manoeuvred their mercenary companies to and fro, avoiding combat where possible to draw payment without incurring losses. Throughout Europe, soldiers were heeding Machiavelli's warning in *Arte della Guerra*, that it was better to tempt fate only when the odds were in your favour. Warfare stagnated. All strategic thinking withered away. War eternalized itself.[34] Then, beginning in the mid 16th century, two northern princes began to change warfare: Maurice, Prince of Orange, and Gustavus Adolphus, the king of Sweden brought forth the so-called Military Revolution of 1560-1660. "[Maurice's] methods were copied throughout Protestant Europe, but not until Gustavus was the reformation truly completed."[35] Predominantly mercenary armies carried out the Mauritian reforms whereas Gustavus Adolphus accomplished his reforms with a conscript militia. However well begun, the changes made by Maurice were not fruitful until Gustavus "for Maurice shared to the full the contemporary dislike of battle. It was left to Gustavus Adolphus to remedy most of the defects of Maurice's system, and in doing so to stereotype European warfare ... until our own day."[36]

Together, the innovations of Maurice and Gustavus were integral to a process that had been slowly taking shape for almost a century. Both men, looking back to Rome, Macedonia, and Carthage for inspiration, made noteworthy changes. The foundation of that change was a return to Roman models in general and to foot-drill in particular:

> Military drill, as developed by Maurice of Nassau and thousands of European drillmasters after him, tapped [a] primitive reservoir of sociality directly. Drill, dull and repetitious though it may seem, readily welded a miscellaneous collection of men, recruited often from the dregs of civil society, into a coherent community, obedient to orders even in extreme situations when life and limb were in obvious and immediate jeopardy.[37]

Maurice replaced the large and unwieldy Spanish infantry *tercios* with smaller and more manoeuvrable units. Gustavus reinvented cavalry, reintroducing shock tactics. The cavalry squadrons, like infantry units, became smaller and thereby more effective, able to exploit any sudden advantage that may have appeared on the battlefield, allowing units to push rapidly into any gap that had been blasted by Swedish artillery or volley musketry. This latter use of artillery was something that Gustavus refined to an art. The use of field guns to break a hole in an enemy line was a tactic that he developed, and which would later be made famous by Napoleon.

More than just the obvious changes of structure and battle posture, the Mauritian reforms, later adopted by Gustavus, also included the critical aspects of digging and discipline. Roman legions and the tactics that won them an empire fascinated Maurice. These tactics, once adapted to a more modern circumstance, were then successfully transplanted into the Dutch army and thence into the Swedish army. "Not only were the Swedish organization, discipline and pay influenced by Maurice's reforms, but the Dutch engineering system was copied almost in detail."[38] As important as the tactical innovations introduced, the era can be said to have seen the birth of logistics. Nations began to retain larger numbers of soldiers even when they were not at war. As a result, administration and regular payment of national armies was initiated with varying success by different monarchs. Uniforms, although by no means new, came to be a more regular practice. The regularization of command-and-control practices created the preconditions, which would someday allow for the creation of a regular and professional officer class.

> [S]ystems of supply emerged during the Thirty Years' War. The first, developed by the Swedes, might seem little removed from plunder but, in that it relied upon a large supply staff, it actually constituted a break from the practices of the period. Gustavus's quartermasters deployed throughout Germany, took inventory of the area's economic resources, and extracted them for the army's needs, without the pillaging, rapine, and massacre normally associated with living off the land.[39]

Although there is little hard evidence that Maurice read Machiavelli, considering the Florentine's wide-spread fame, it is difficult to believe that he was not influenced by Machiavelli's *Art of War*, *il Principe* and *Arte della Guerra*. It is a legitimate criticism that Machiavelli underestimated the effects of technological change. None-the-less, his political and moral lessons were widely accepted, and considering his model of military excellence was the Roman legion, it would be difficult to accept that Maurice was not influenced by his writings.

At any rate, Maurice made some dramatic changes:

> The improved efficiency aroused by drill soon became apparent to other military men of Europe. Prince Maurice's reputation rested on his recovery of dozens of fortified towns from the Spaniards through sudden strikes and obdurate sieges, each conducted with a technical precision and dispatch never attained before. Maurice's methods of training were not kept secret...
>
> Maurice organized a military academy for the training of officers in 1619 – another first for Europe. A graduate of Prince Maurice's academy subsequently took service with Gustav Adolf of Sweden and brought the new Dutch drill to that army. From the Swedes the new drill (variously modified of course) spread to all other European armies with any pretension

to efficiency.[40]

The smaller 'reinvented' cavalry units once again became a commanding presence on the battlefield. Likewise, the smaller infantry units were once more flexible and more potent. Linear tactics were rediscovered. Artillery was developed rapidly and given an ever-growing importance in battle. The science of fortification was both rediscovered and reinvented with the introduction and spread of the *la trace italienne*. Armies grew. The age saw most of Europe become an armed camp with larger and larger armies either investing each other's fortifications or manoeuvring against each other.

Together, these two men shook warfare from its medieval torpor. The reduction of the size of tactical units, the integration of artillery as an early and coordinated member of the triumvirate of combat arms, the reintroduction of drill and the creation of military academies for the scientific study of warfare all contributed to making armies more efficient, more deadly, and more useful. The modern concept of logistics and early attempts at standardization began to move warfare away from being what Martin van Creveld described as little more than a 'walking tour combined with large scale robbery.' Gustavus has sometimes been given credit as the Father of Modern Warfare, but his innovations could not have occurred had it not been for Maurice. The changes introduced by Maurice, improved upon by Gustavus and copied by future commanders changed war and warfare forever.

Maurice Marshal de Saxe (1696-1750)

German born, but a Marshal of France, de Saxe was arguably the most revered soldier of his day. He was the eldest of more than reputedly 350 illegitimate children of Augustus II, Elector of Saxony, and King of Poland. Commissioned at age twelve, a regimental commander of cavalry at seventeen, de Saxe focused on the human side of war. Part soldier, part tactician, part *philosophe*, he was convinced that the secret to mastering the art of war lay in the human heart. Although de Saxe emphasised the human aspect of warfare he did not go so far as to embrace the thoughts of the English philosopher John Locke, a strong proponent of Thomas Hobbes' social contract. As his most famous subordinate Lloyd would later write, he felt that commanders should develop those qualities they found in soldiers rather than attempt to make soldiers something that they were not: "that military discipline, tactics, and training should complement rather than counter human nature."[41]

Saxe did not believe that war was a science, even if he called it such: "War is a science covered with shadows in whose obscurity one cannot move with an assured step. Routine and prejudice, the natural result of ignorance, are its foundation and support. ... All sciences have principles and rules; war has none."[42] Rather than principles, he saw that there were what are now referred to as tactics, techniques and procedures, which, if followed, could gain success. His book is filled with such tips to commanders as not to be too harsh in punishing soldiers, the use of vinegar to reduce disease from bad water and that the secret of successful battlefield manoeuvre was efficient marching. The profession of arms, he said, had little to do with arms but rather with the legs, that is, train troops to march. By the time de Saxe came to the fore, the rediscovery of Roman drill and Roman discipline had already had its impact, but de Saxe went further, developing the art of cadence marching:

> Considering what I have said superficially [about men marching to music], it does not appear that this cadence is of such great importance. But, in a battle to be able to augment the rapidity of march, or to diminish it, has infinite consequences. The military step of the Romans was nothing else; with it they marched twenty-four miles in five hours. ... Among the Romans this was the principal part of their drill. From this ... one can judge the importance of cadence.[43]

The reason that he placed such emphasis on manoeuvre was that warfare had stagnated due to the advancement of fortifications by Vauban and the wide use of *la trace italienne*, a style of fortification developed in Italy in the late 15th and early 16th centuries in response, primarily, to the French invasions. Another technique developed by him in this respect was the use of irregulars in front of his formed battalions. Arguably, his model is the earliest record of the consistent use of these troops, which would eventually be formed in special units and come to be called *tirailleurs*. Like Maurice and Gustavus before him, he believed that infantry battalions had grown too large and strove to break them into smaller entities. Saxe concluded that it was not large armies that won battles – it was good ones.

Saxe was influential during his life due to his great battlefield successes, but even more so after his death with the publication of his *Reveries on the Art of War*. The book became required reading for any European who fancied himself a soldier. It is a poorly organized, rambling and often inconsistent tactical handbook for would-be commanders. Nonetheless, it is filled with good advice. Saxe's combination of innate tactical brilliance, charisma, and innovations such as cadence marching and the use of skirmishers, made him a target of emulation and his book remains required reading to the present day. Although referring to warfare as a science, it was clear that the marshal considered it to be an art – albeit an art that could be improved upon when studied.

Frederick the Great (1712-1786)

The father of modern Germany and idol of Napoleon initially showed no interest in the military arts, much to his father's disgust. His father court-martialled and imprisoned his Crown Prince, forcing him to work with the army as punishment. Upon his succession to the Prussian throne, he discovered an innate talent for tactics as well as for strategy. From his succession in 1740, to his death, he more than doubled the size of his kingdom and almost trebled its population – mostly by military conquest and acquisition.

Frederick was passionately interested in the arts as well as in war. A true philosopher-warrior-king, he was a friend and correspondent of Voltaire. In warfare, he was influenced by one of France's authorities on the art of war, Antoine de Pas, *marquis* de Feuquières (1648-1711), who strongly stressed the need to conduct war aggressively and audaciously as a response to the methodical scientific precision of Vauban's (1633-1707) siege warfare. Although harshly judged by history for his brutal iron discipline, Frederick was a great king as well as a successful battlefield commander. Building upon the advances of mobility as described by de Saxe, he turned battlefield drill into a precise science. His use of the oblique attack was impossible without highly drilled troops. At the Battle of Leuthen in 1757, its use not only destroyed a larger Austrian army; it set Frederick apart as a tactical genius.

Arguably, Frederick is the first example of a professional commander. He worked his soldiers incessantly to increase both their skill and their speed of execution: "You will have seen by what I have had occasion to delineate concerning war that promptness contributes a great deal to success in marches and even more in battles. That is why our army is drilled in such fashion that it acts faster than others."[44] Frederick's *Instructions* could be summed up in two words: study and offence. His *Instructions* are consistent in their admonition to study and plan, to prepare and once ready, to avoid passivity:

> Let no one imagine that it is sufficient just to move an army about to make the enemy regulate himself on your movements. A general who has too presumptuous confidence in his skill runs the risk of being grossly duped. A great deal of knowledge, study, and meditation is necessary to conduct it well, and when blows are planned whoever contrives them with the greatest appreciation of their consequences will have a great advantage. However, to give a few rules on such a delicate matter, I would say that in general the first of two army commanders who adopts an offensive attitude almost always reduces his rival to the defensive and makes him regulate himself on his movements.[45]

The Prussian king was the master of the battlefield during his lifetime and his *Instructions* a highly prized secret. A copy was captured in 1760, translated and quickly printed. It soon became required reading for all officers in Europe who were serious about their profession. Scharnhorst, for instance, used it in his *Militärische Gesellschaft* and Napoleon claimed to have studied it as he studied no other text. Frederick's battlefield successes, as well as his writings, entered the military canon during his lifetime and have been studied ever since. More than two centuries after his death, the Prussian king who initially had no interest in things martial is still studied at the major

European military colleges and by extension his influence continues to be felt by most Western armies.

Napoleon (1769-1821)

Although Napoleon did not leave behind any writings or cogent military theory *per se* his influence upon modern war cannot be overstated. He was the undoubted master of his battlespace. Both the magnitude and the brilliance of Napoleon's campaigns unquestionably changed both the conduct and the understanding of war in the Western world. His conquest of Europe forced a watershed in military thought for as Professor Peter Paret explains, his military successes launched a broad search to find the underlying explanations for his astounding battlefield successes. His armies demonstrated innovations, which many opponents had not previously encountered. French infantry was more flexible in its employment. So was the artillery. Troops moved in self-contained divisions, groupings that allowed better coordination among the various combat arms. The supply system was less rigid. The idea of privilege in the selection of officers was abandoned. Mass conscription swelled the size of his armies. The cautious tactic of choosing and holding key points was replaced by rapid dispersed marching culminating in breathtaking concentrations of troops against opponents in the field.

Whether Napoleon was personally responsible for the many changes that he brought to the art and science of war is questionable. If we consider the massing of troops at decisive points, the innovative use of *tirailleurs* or the integration of artillery with infantry and cavalry, all these individual tactics and techniques had been seen before. Nevertheless, credit must be given to him for the way he combined all these available tools:

> Napoleon was not himself a reformer; with a profound understanding of their potential, he made use of forces that had already been created. Earlier commanders might also have dreamt of strategies that sought decision in climactic battles. So long as they led armies of expensive mercenaries whose reliability could be assured only by stringent control and care, they could not cut loose from their supply bases. They were compelled to fritter and fragment their troops in the defence of every position and to limit the risk of battle. In the revolutionary and imperial armies, however, much more could be demanded of the soldier. Soldiers were now more expendable, which rendered the risk of battle less onerous.[46]

The French Revolution and its subsequent wars changed the political, and with it, the military situation not only in France but in all Europe. The fact that Napoleon had not invented the concept of divisions (de Saxe), or the need for extreme rapidity

(Frederick) does not diminish his battlefield brilliance. He saw possibilities in a way that others around him did not; he seized upon the opportunities as they presented themselves.

Part of his genius was to have his opponents believe that he was a genius, that other commanders could not understand his methods of warfare and Professor David Chandler explains how the Corsican worked to build his own image:

> Napoleon was only too pleased to have his contemporaries believe him to be a unique military phenomenon – and through his propaganda set out sedulously to foster the image, both during his period of power and during the last years on St Helena. His main means to this end was steadfastly not to explain his methods in any great detail, or even in outline, thus fostering an illusion of sublime and unique abilities incapable of comprehension by ordinary mortals.[47]

There has never been any lack of interpreters of Napoleon's brilliance. They range from his contemporaries and subordinates like Clausewitz and Jomini, to modern writers who ascribe Bonaparte's success to "French battlespace dominance" that led to "strategically focused, sequential operations and engagements culminating in a dominating maneuver to destroy the enemy's armed might."[48] But more important than all the volumes written on how Napoleon was able to control time and space was the effect that he had upon one of his enemy's reforms. A small group of Prussian officers, intent on reform, captured the essence of the Napoleonic leadership style and was able to turn this style into an institutional manifestation of how to lead men in battle. Beyond the aspects of tactical victory and political conquest, Napoleon, like no European general before him, set a new modern standard for battlefield excellence.

No soldier-statesman in the modern era has had as great an effect upon the conduct of war as Napoleon; he was the first great strategist of the modern age. Like a catalyst that allows a chemical reaction to occur, his actions changed modern warfare. His influence was both pervasive and enduring. Whether at the tactical level with the adoption of techniques he used or at the operational level and the search, even in the late 20th century, for 'Napoleonic' battles or at the theoretical level where entire schools of military thought have grown up around attempts to explain his success, no individual has had a greater impact upon warfare in the modern age.

Helmuth von Moltke (1800-1891) and Alfred von Schlieffen (1833-1913)

It is impossible to think of modern European warfare without conjuring up images of two Prusso-German Chiefs of the Great General Staff (GGS). Together, they formed the basis for the professionalism and excellence of their nation's army. Of the two,

Moltke is undoubtedly more responsible for shaping German military thought in the 20th century:

> By the end of his long career, the elder Moltke had significantly increased the capabilities, influence, and prestige of the GGS [Great General Staff], developing and refining the staff-ride system, demanding the highest standards of professionalism from GGS officers, gaining 'direct access' (*Immediatvortrag*) to the Kaiser in 1871, and obtaining direct control over the training program at the War Academy in 1872. In fact, the Prussian army and its general staff system had become the envy of and model for other armed forces both on the continent and beyond.[49]

Clausewitz had been the commandant of the Prussian staff college when Moltke was a student and although he later broke with some of Clausewitz's teachings, Moltke often cited him as a primary influence upon his professional development. Moltke's thoughts on strategy and tactics stressed flexibility of both thought and action. From him comes the admonition that a plan never survives first contact with the enemy (*Ein Plan hält nur, bis zur ersten Feindberührung*), as well as the need to give commanders at all levels freedom of action (*Freiheit des Handelns*). Further, it was Moltke who trained the Prussian army to work with minimalist orders, believing that commanders would be well served only to order that which was 'absolutely necessary'. But Moltke differed from Clausewitz in one critical aspect. He believed it necessary to separate the military operation from the political realm. His argument with Bismarck after the crushing Prussian defeat over Austria at Königgrätz is a textbook example: Moltke wanted to annihilate the Austrian army whereas Bismarck knew that he would gain political advantage from allowing the Austrians a negotiated peace. In other words, use the tactical victory for political ends, even if the tactical victory needed to be reined in. The crux lay in the fact that the core of Moltke's military thinking revolved around tactical victory.

Moltke spent an incredible thirty-one years as Chief of Staff and his life's work, and the legacy that he gave the GGS, was the nettlesome problem of a war on two fronts. He worked tirelessly on the problem, constantly modifying the plans, and updating them as technology altered warfare. GGS officers were to plan ceaselessly, to be faceless and nameless, to strive for ultimate flexibility and never to tie the hands of battlefield commanders (*Auftragstaktik*).

Barely three years after Moltke ended his astounding tenure as Chief of Staff, Schlieffen took the reins. Moltke's great strategic concern, an attack against Germany on two fronts, soon became Schlieffen's great passion. Significantly influenced by both Clausewitz and Moltke, the new chief drew upon his historical studies of the Roman battle of Cannae (216 BC) and thus gave the German army the concept of the battle of annihilation. Schlieffen's belief was that tactical victory lay in the flank attack and his thinking was soon imbedded in German tactical thought for the remainder of the 20th century. Schlieffen's other contribution is a negative lesson: his misinterpretation of Clausewitz combined with the obsession with winning the two-front-war, combined with the continuation of Moltke's strong separation of the military and political spheres, led Germany into a war that, through the violation of Belgium' neutrality in 1914, soon engulfed all of Europe. Thus, although best remembered for his ill-fated plan, Schlieffen's real legacy lies in the inculcation of the 'perfect' tactical technique of the flank attack combined with the admonition that Clausewitz's dictum of war and politics cannot be set aside in favour of purely military consideration, for the civil-military relationship is ignored at great peril to the nation.

Thus, the legacy of Germany's two best 19th century Chiefs of Staff was to create and maintain a highly professional general staff, and thereby, in Dupuy's words, to 'institutionalize military excellence' in Germany. Moltke's model was so successful that it was copied all over the continent, in the United States, as well as in Japan. Schlieffen's emphases upon the attack against a flank and the battle of annihilation entered the psyche of the German military and have endured for more than a century.

Second World War Manoeuvrists

The Second World War was the first war of fully mechanized manoeuvre. When the Wehrmacht dashed into Poland in 1939, and repeated its performance against France in 1940, the stage was set for all armies to change the way that they perceived and fought war. The promise of the great 'breakout battles' as foretold by Fuller, and Liddell Hart had come to pass. General Heinz Guderian's audacious dash from the French border to the English Channel seemed almost unreal. Despite having more divisions, better weapons and the 'impenetrable' Maginot Line, the French army collapsed in mere weeks. But it was not a failure of fighting. It was a failure of doctrine that was based upon flawed military thinking that spawned a defective theory of defensive warfare. In his memoirs, *Achtung Panzer!* Guderian gives credit to the German army's willingness to absorb and internalize the theories of Fuller and Liddell Hart. Combined with a heritage of flexible leadership and the professionalism of the famous German General Staff a new (if flawed) form of warfare emerged: *Blitzkrieg*.

Blitzkrieg

Blitzkrieg has long captured the military as well as the popular imagination. This new development in warfare was not, however, a planned war-winning strategy. In some ways it was an accidental outcome of years of study coupled with a military theory based on independent thought and action. Further, it was not a strategy at all, although it was used as one. With its intellectual roots in the theories of Clausewitz and its operational roots in the thinking of Moltke, Schlieffen, and the campaigns of Frederick the Great, it took on the appearance of a strategy, without the necessary components thereof. What made *Blitzkrieg* unique was that it, in effect, turned conventional wisdom on its head. Among other innovations, tanks were released from their traditional infantry support role; they were grouped in armoured formations, which were fully motorized and independent, both tactically and logistically. Although they had their own artillery, the *Luftwaffe* was also assigned directly in their tactical support. Armour was tasked to operate at a high tempo that could not be sustained either by the enemy or by friendly units. These new weapons were not used to attack defensive positions, they were assigned to circumvent them and to look for reserves to destroy.

So, if it was it not a strategy, then why was it so successful? Between 1936 and 1942, Nazi Germany was almost unfailingly successful in battle. Adolf Hitler made a series of bold and dramatic moves, both political and military. But, late in 1941, the tide turned. Why? Why was Germany so successful for so long but unable to hold on to its gains? Of course, there is no simple explanation to such a multi-faceted issue. But this question offers an opportunity to demonstrate the difference between getting all the facts straight and still not understanding what happened.

First, let us look at the reason for German success between 1936 and 1942: Germany, because of Hitler's ability to divide potential opponents, fought a series of campaigns rather than a war, campaigns that were separated from one another by time and space and campaigns that were mostly directed against individual states massively inferior to Germany whether in terms of population, industry, military power, or geography.

The German army had spent the interwar years under the tutelage of General Hans von Seeckt as the head of the *Heeresamt* or Army Office. Seeckt was a cerebral General Staff Officer of the 'old school', and he built this organization, which was the renamed, outlawed *Grosse General Stab*. The *Heeresamt* spent the two decades between the wars scrubbing German doctrine and studying the lessons learned from Germany's loss in 1918. Versailles may have limited the rump of the Imperial Army, the *Reichswehr*, to 100,000 men but most of them were officers and NCOs or at least trained to rise to those ranks quickly. They became the seeds of the future multi-million-man Wehrmacht that would at its height approach 10,000,000 personnel.

When the Nazis took power, the army's leadership was wary of entering another long, drawn-out war. Worse, they dreaded the possibility of a two-front war (recall Moltke and Schlieffen). Hitler's political maneuverings reinforced their fears. The officers trained by Seeckt saw that Hitler was creating national preconditions for just such a prolonged bloodletting; something which all the General Staff officers who had fought the last war were eager to avoid. But each short sharp campaign, seemed to invalidate their fears. All of Hitler's targets fell like dominoes. But Germany's tactical and operational successes up to 1942 taught the Wehrmacht false lessons. The short, violent, and rapid campaigns lulled both the military and the political leadership into believing that no country was a match for Germany's might. The difficulty with this successful pseudo-strategy was that no operational or strategic concepts underpinned it, a fact that became increasingly obvious on the Eastern Front, in part because the size of the theater demanded an operational level of war, in part because of the fundamental German error of seeking victory in a single decisive 'Napoleonic' battle, and in part because tactical concentration could not produce operational victories. In short, it was *not* a strategy.

The hubris that was born of the victories of 1936-42 was ready to blossom once the US entered the war. General George C. Marshall correctly foresaw that a long war strategy was the only way that America could hope to win a war that was to be fought on a global scale. America's strength was her economic potential and so he did what every great strategist has done: he forced his enemies to play by his rules. At first, the American strategy did not appear to work but as with all long-game strategies, once it took hold, it was a race that only one horse had the legs to win.

Instead of looking at it from strategic, operational, or tactical perspectives, let us look at it conceptually. Let us reconsider the model I suggested above, at the level *above* strategic. Look at the situation theoretically. Professor Antulio Echevarria II has elegantly described the situation thus: Germany confused *force* with *power*. German leadership, principally the senior Nazi officials, confused the country's military force with national power. (This was an easy mistake to make and in fact it is one that Americans made several times in the 20th century). The Nazis failed to comprehend the limitations of force within the greater international community.

In effect, what the West knew as *Blitzkrieg* was the physical manifestation of this misunderstanding. This form of warfare was a tactical concept, but Hitler mistook it for a strategy. It might well have been useful as such except that the Nazis did not configure the armed forces to carry out such as strategy. The myth created by *Blitzkrieg* contained the seeds of its own destruction. The more they tried to use it, the worse it became because the short sharp engagements used up resources that Germany could ill afford whereas the Americans and the Soviets could absorb the short-term losses if the long-term strategy made progress. The Allied strategy had many parallels with that which Ulysses Grant eventually used to defeat Robert E. Lee. Grind your opponent down.

As an aside, there is an excellent example of this error in the publishing industry: From 1883 to 2000, one of the world's premier magazines was *Life Magazine*. During its golden age in the mid 20th century, it was the gold standard for high quality photojournalism. But starting in the late 1990s it started to lose money and at the end, it cost more to publish each issue than the publisher made on each copy. The more copies they sold, the worse the financial situation became. The company's success killed the magazine.

To summarize, the Germans inadvertently got themselves into a long war but only had a blueprint that allowed them to fight a succession of short wars (or more properly, campaigns) – most of which they won. But to paraphrase Vietnamese Colonel Tu when he described why the Americans lost the Vietnam War, the fact that the Germans continually outfought the Soviets and the Allies at the tactical level eventually became irrelevant. The Germans had forgotten their Clausewitz.

But the Germans were not the only innovators during the Second World War. Arguably, the greatest tactical commander on the Allied side was American general George S. Patton Jr. (1885-1945). Patton was the first US officer trained as a tank commander and an early and ardent proponent of mechanized warfare. He was also an early pilot. Long a student of warfare, he was a voracious reader, owning personal copies of all the great military classics. Studying the German victories, he quickly came to appreciate the necessity of unleashing the power of tanks from their subordinate role as merely support to infantry. Although not necessarily an innovator, Patton quickly demonstrated that the German techniques could be copied and even improved upon. Combining such tactical brilliance with America's overwhelming materiel dominance, and supporting it with air superiority, generals like Dwight Eisenhower (1890-1969) and Douglas MacArthur (1880-1964) proved that no strategy was perfect, and that any technique could be enhanced.

A short word on both Erwin Rommel (1891-1944) and Erich von Manstein (1887-1973) is in order. Although both generals were primarily combat officers, the former being more of a tactician and the latter more of a strategist, their influences continue to be felt to this day. Rommel's, 1937 *Infanterie Greift an: Erlebnisse und Erfahrungen* (*Infantry Attacks*) continues to be read by infantry officers around the world. Even if the book is based on his experiences during the First World War, it remains a valuable manifestation of the asymmetric thinking that made German tactical fighting so successful, and so emulated. In a similar vein, but at a higher level, Manstein's 1955 *Verlorene Siege* (*Lost Victories*) is pure Clausewitz. Despite the taint of war crimes, Manstein was part of the advisory group that helped to create the Bundeswehr in 1955 and his writings, as well as his operations, continue to be studied at the *Führungsakademie*.

Summary and Conclusions

In this chapter we have surveyed of the work of the great minds of war on land. Their collective works form most of the body of Western military thought and theory. An in-depth study was not possible here; but neither was that the intent. The chapter has been selective of necessity. The purpose was to demonstrate two key aspects of the study

WAR ON LAND

of war on land: First, that there is a necessary difference among philosophers of war, military theorists, and military strategists. They are distinct and not interchangeable. The theorist builds his ideas upon those of the philosopher; and the strategist develops his thoughts based upon those of the theorist. Second, that even though Western military thought and theory is an intricate and growing matrix of thoughts, beliefs, and practices, most of the men studied here repeatedly tilled the same ground. Rarely was an idea, concept, or action truly new. More often, the idea was an improvement of an earlier premise, the perfection of a previous action, a better explanation of an old idea. Over the course of more than two millennia, there has been much confusion among what is philosophy, what is theory and what is strategy. There has been more recycling than creation.

Humankind has practised the art and science of war on land for millennia. One thing has remained constant in all this time: Whether discussing Alexander's victories over the Persians or the US army's 21st century attack on Iraq, battles have always been confused mêlées of death and destruction. Understandably, therefore, the intellectual history of land warfare has represented the desire to make sense of the confusion, to bring order to what is inherently disorderly. Whether you considered the dilemma philosophically, as did Sun Tzu, attempted to create a theory that brought discipline to thinking like Clausewitz and Jomini or simply attempted to form strategies to gain military and political objectives, the quest has been the same: the pursuit of military victory.

This pursuit has been the military equivalent of the search for the Holy Grail. From believing in supernatural intervention in classic antiquity, to depending on military genius, to believing in the power of technology, this pursuit has failed to provide a permanent and definitive solution. However, this lack of an ultimate solution should not be seen as failure. Although the search has not produced any definitive theory of war, it has kept humankind aware of the need for constant study. More than in any other human activity, stagnation, complacency, and ignorance of war can result in dire consequences. If nothing else, the continual pursuit of insights into the nature of war has been both useful and beneficial.

The ancients understood war as a natural condition. It was a common social behaviour. Possession of a theory or a collection of principles was unimportant to them compared with the practical knowledge required to fight. Thus, rather than actual theories, their appreciation for war was collected wisdom regarding tactics, structure, administration, and training. From a conceptual perspective, classic Greek and Roman military theories were either religious or so practical as to be little more than field manuals. This collection of techniques was the legacy left behind when the Roman Empire collapsed. In the centuries following the fall of Rome new ideas occasionally emerged but as John I. Alger points out in his study of the principles of war, *The Quest for Victory*, military schools became the focal point for thought and reflection on war. Over time, the growing needs for comprehensive texts on warfare, strategy and tactics stimulated some to write. Others published their lectures and notebooks. Military treatises proliferated. Along with them came various schools of thought. In this way, interconnectedness grew among most of Western society's military philosophies, theories, and strategies. Themes recurred; ideas were borrowed; influence and nuance were carried over from one person to another and from one generation to the next.

As discussed, philosophy forms the foundations of both theory and practice. One of the distinctions between philosophy and theory is the timelessness of the former compared to the latter. When we consider the writings of Sun Tzu, it is apparent that his work owes much of its longevity to his lack of technical advice. His words are not temporally dependant. This ageless quality has not been present in the works of most of those who followed Sun Tzu, with Machiavelli being a notable exception. Conversely, most military theorists have based their works on specific technologies that were characteristic of their own times. An exception might be argued for Clausewitz's. Although clearly grounded in the author's experiences with late 18th and early 19th century conflict, *On War* frequently transcends this experience to grasp some of the cerebral components of war. Wherever an individual has made such an attempt – to elevate his investigation of war above the mundane – inevitably his thoughts and words have endured.

The modern Western way of war is strongly connected to its past. Sometimes the connection is linear and easily traceable backwards along the arrow of time; sometimes not. Either way, the past informs the present and the future. Beginning with the ancients' understanding of the nature of war, society continually sought to improve its collective understanding of war. Rarely, new philosophies, theories and strategies did bring such understanding. Machiavelli brought a better understanding of the connectivity between political power and military arts and sciences. Clausewitz and Jomini each, in their own ways, helped observers appreciate the strategies and tactics of Napoleon. Sometimes the viewpoint, as with de Saxe, stressed the human aspects of the art of war. Sometimes, as with Lloyd, it sought to uncover some underlying rules that a commander could follow. Whatever the view, the tool was inevitably the same: historical study. Looking backwards helped provide the guideposts to move forward.

Having briefly reviewed the works of prominent and influential philosophers, theorists and strategists, the relationship among this 'remarkable trinity' should now be clearer. The philosopher forms the bedrock of theory. The theorist creates a paradigm for a belief system; a supposition or system of ideas explaining something, especially one based on general principles, like the exposition of the principles of a science or the collection of propositions to illustrate the principles of a subject. Last, the strategist translates theory into practice to gain some objective. The boundaries that divide these characters are not always clearly drawn; sometimes an individual could be included in any or all the groups. No matter. The key is understanding is the appreciation that a relationship exists among the three components and to distinguish one component from another.

WAR ON LAND

The manner of fighting war on land has been a story of change but, ironically, the nature of the struggle has remained essentially the same. A single recurring theme should be obvious. Whether you study Sun Tzu, or Machiavelli, whether a student of Clausewitz or Jomini, whether preferring the strategies of Frederick the Great or those of Manstein, the nature of war remains unbreakably connected to human nature. In other words, war remains a human endeavour. Irrespective of the new machines that humankind has brought to the battlefield, humans instigate, create, and conduct wars. Remembering this fact, the lessons of the past remain invaluable. Whether you look back to the period of the Warring States in China for Sun Tzu's insights or merely to the recent past of the Second World War, there are insights to be gleaned from the thoughts and actions of the past.

Notes

1. Victor Davis Hanson, "The Western Way of War" Australian Army Journal, (Winter 2004), 159.
2. John Alger, *The Quest for Victory, The History of the Principles of War* (Westport, CT: Greenwood Press, 1982), 179.
3. Sun Tzu, *The Art of War*, Samuel B. Griffith, translator, (Oxford: Clarendon Press, 1963), I:1.
4. Michael Handel, *Masters of War: Classical Strategic Thought*, (New York: Frank Cass, 2001), 28.
5. Alger, 7.
6. Niccolò Machiavelli, *Il Principe*, (Milano: RCS Rizzoli Libri, 1986), XXVI: 3. Machiavelli is quoting Titus Livus, Histories, IX:1. (My translation).
7. Alger, 5.
8. Peter Faber. "Strategy: Its Intellectual Roots and Its Current Status", excerpted from "The Evolution of Airpower Theory in the United States – From World War I to John Warden's The Air Campaign," Asymmetric Warfare, John Andreas Olsen ed., Norwegian Air Force Academy Militæteortisk Skriftserie No. 4, 2002, 2.
9. Patrick J Speelman, *Henry Lloyd and the Military Enlightenment of Eighteenth Century Europe*, (Westport, CT: Greenwood Press, 2002), 16.
10. Gunther Rothenberg, "Maurice of Nassau, Gustavus Adolphus, Raimondo Montecuccoli, and the 'Military Revolution' of the Seventeenth Century", Makers of Modern Strategy: from Machiavelli to the Nuclear Age, (Princeton, NJ: Princeton University Press, 1986), 63.
11. Henry Lloyd, *The History of the Late War in Germany between the King of Prussia and the Empress of Germany and her Allies, Part I Vol 2*, (London: S. Hooper, 1781), xii. (Italics in original.) This same idea is found in de Saxe and undoubtedly borrowed from Vegetius.
12. Henry Lloyd, *The History of the Late War in Germany between the King of Prussia and the Empress of Germany and her Allies*. (London: S. Hooper, 1766), 6.
13. Henry Lloyd, *The History of the Late War in Germany between the King of Prussia and the Empress of Germany and her Allies, Part I Vol 1.*, (London, S. Hooper, 1781), 29.
14. Henry Lloyd, *Preface to Continuation of the History of the Late War in Germany between*

the King of Prussia and the Empress of Germany and her Allies Part II. (London, S. Hooper), i.
15. Michael Howard, "Jomini and the Classical Tradition in Military Thought", in Michael Howard, ed., *The Theory and Practice of War*, (Bloomington, IN: Indiana University Press 1975), 9.
16. Paul D. Mageli and David Bongard, *Harper Encyclopedia of Military Biography*, (New York, NY: Harper Collins, 1992), 113.
17. Walter Görlitz, *History of the German General Staff 1657-1945*, Brian Battershaw translator, (New York, NY: Praeger, 1953), 33.
18. Charles Edward White, *The Enlightened Soldier: Scharnhorst and the Militärische Gesellschaft in Berlin, 1801-1805* (Westport, CT: Praeger, 1989), 5-6.
19. Ibid, 16.
20. Ryan Henry and C. Edward Peartree, "Military Theory and Information Warfare", Parameters, (Autumn 1998), 122.
21. Thomas M. Huber, *The History Of Warfighting: Theory And Practice, Intro to Lesson 8: Interpreting Modern War - Jomini*, U.S. Army Command And General Staff College. available at http://www.au.af.mil/au/awc/awcgate/jomini/lsn08.htm. Note: this is an archived page and must be accessed using an Internet Archive service.).
22. Peter Paret, *Innovation and Reform in Warfare, The Harmon Memorial Lectures in Military History*, No. 8 (Colorado Springs, CO: United States Air Force Academy, 1966), 8.
23. Martin Van Creveld, *The Transformation of War* (New York, NY: Simon & Schuster, 1991), 34.
24. Robert P. Pellegrini, *The Links between Science, Philosophy, and Military Theory, Monograph, School of Advanced Airpower Studies*, (Maxwell Air Force Base, AB, 1997), 17.
25. Wager Halleck, *Elements of Military Art and Science; or Course of Instruction in Strategy, Fortification, Tactics of Battles, &c; Embracing the Duties of Staff, Infantry, Cavalry, Artillery and Engineers*, 1971 reprint of 1846 Appleton edition, N.Y. (Westport: CT, Greenwood Press, 1971, 37. (Emphasis in the original.)
26. Ibid, 48-58.
27. Ibid, 49.
28. Russell F. Weigley, *The American Way of War: A History of U.S. Military Strategy and Policy* (Bloomington, IN: Indiana University Press, 19840, 98.
29. Basil Liddell Hart, Strategy, 2nd ed., (New York, NY: Praeger, 1968), 212.
30. J.F.C. Fuller, *Foundations of the Science of War*, (London: Hutchinson and Company, 1926), 47.
31. Ibid, 314.
32. From Palmer McGrew's "Experience in Vietnam with A Company Commander, de Czege" available at http://www.west-point.org/users/usma1958/22097/3tactics.html.
33. Huba Wass de Czege, "The Continuing Necessity of Ground Combat in Modern War," Army, (September 2000), 2.
34. Michael Roberts, *The Military Revolution 1560-1660*, (Belfast: Queen's University Press, 1955), 7.
35. Philip Haythornwaite, *Invincible Generals: Gustavus Adolphus, Marlborough, Frederick the Great, George Washington, Wellington*, (Bloomington, IN: University of Indiana Press, 1991), 15.
36. Roberts, 8.

37. William H. McNeill, *The Pursuit of Power; Technology, Armed Force, and Society since AD 1000*, (Chicago: University of Chicago Press, 1982), 131.
38. Lynn Montross, *War Through the Ages*, (New York, NY: Harper & Brothers, 1944), 269.
39. Brian M. Downing, *The Military Revolution and Political Change*, (Princeton, NJ: Princeton University Press, 1992), 71.
40. McNeill, 134.
41. Speelman, 26.
42. Hermann Maurice, Comte de Saxe, *Reveries on the Art of War*, Thomas R. Phillips, translator and editor, Roots of Strategy edition, (Harrisburg, PA: Military Service Publishing, 1955), 189.
43. Ibid, 205
44. Frederick the Great, *Instruction of Frederick The Great for His Generals*, 1747, Thomas R. Phillips, translator and editor, Roots of Strategy edition, (Harrisburg, PA: Military Service Publishing, 1955), 394-395.
45. Ibid, 363.
46. Paret, Innovation and Reform, 5.
47. David Chandler, *Napoleon*, (London: Weidenfeld and Nicolson, 1973), 184-185.
48. Douglas A. MacGregor, *Breaking the Phalanx*, (Westport, CN: Praeger, 1997), 40.
49. Antulio J. Echevarria II, "Moltke and the German Military Tradition: His Theories and Legacies," Parameters (Spring 1996), 91.

" Since men live upon the land and not upon the sea, great issues between nations at war have always been decided – except in the rarest cases – either by what your army can do against your enemy's territory and national life, or else by the fear of what the fleet makes it possible for your army to do.

Julian Corbett
Some Principles of Maritime Strategy

WAR AT SEA

Introduction

We now turn to naval theorists and strategists. Based upon one or other philosophies, the men discussed below created theories that allowed humankind to better understand sea power. For reasons previously enunciated, we will focus primarily upon the theorists Alfred Mahan and Julian Corbett. Other theories and strategies are only discussed to round out the picture of war at sea. The chapter's value is in gaining an appreciation of both the influence of land theorists upon the development of war at sea as well as the influence that naval theorists may have had upon gaining an insight into the nature of war in general.

The essence of naval theory is straightforward: all navies are based on the premise of having or denying freedom of the seas. More precisely, navies offer a state the ability to protect its use of the sea as a means of communication while simultaneously denying the same to an adversary. From a theoretical perspective, the basis of sea power can be expressed in terms of the difference between those theories that propose the primacy of 'command of the sea' and those that profess 'control of the sea.' The former school developed from the writings of Admiral Mahan whereas the latter was the intellectual offspring of Corbett. Although both authors used the expression 'command of the sea', as we shall see their intents were different and it was this difference, which became contentious. The tension between these two schools came to define the great debate of naval theory, from its inception at the turn of the 20th century to the present day.

Unlike armies, navies offer the ability to use the sea to project political power to a foreign shore – whether the shore of a friend or a foe. This ability is a major distinction between navies and armies. Navies have peacetime political uses, which are almost as important as those of wartime: they can project power; they can be instruments of foreign policy; they can be used to promote trade:

> In wartime, the political uses of sea power are naturally relegated to the background in the formulation of naval strategy, which concentrates on combat capabilities, i.e., 'sea control' and 'projection,' ... In the absence of general hostilities, however, a reverse priority applies, and though the prolonged confrontation of the Cold War has retarded the process, the focus of Great Power naval strategy has been shifting to missions that are 'political' in the sense that their workings rely on the reactions of others, and these are reactions that naval deployments may evoke, but cannot directly induce.[1]

Over the millennia, the desire for command of the sea has remained the foundation upon which all maritime nations have built their naval strategies. The fact that the Romans referred to the Mediterranean as *Mare Nostrum* (Our Sea) is indicative of this concept. At its heart, this is the thesis that Mahan espoused in 1890 in *The Influence of Sea Power upon History*. Whether with triremes, barques, men o' war, dreadnoughts or nuclear-powered aircraft carriers, command of the sea has remained a fundamental objective of all naval strategy. In the words of British Admiral Sir Herbert Richmond, the aim of all naval strategy has always been to "render the sea a safe road for one's own movements, a dangerous road for the movements of the enemy."[2]

Whether a nation seeks command of the sea, or only control of it, it obviously requires navies; but unlike armies, which at least from a theoretical perspective can come in many guises, there are only two types of navies: *blue water* and *coastal*. Blue water navies, as the term implies, sail the high seas. They comprise armed ships that can be sent around the world to project power to any corner of the globe, to threaten any shore; their sailors have the skills and knowledge to fight engagements with other blue water fleets hundreds or even thousands of miles from their home ports. Coastal fleets, on the other hand, do not have the ships, the mandate, the skills, or the tradition to sail the high seas; they are coastal protection forces that are designed, trained, and equipped to protect sovereignty locally, to intercept intruders, to patrol coastal waterways in support of friendly merchant shipping, or to destroy enemy merchant shipping. Whereas a blue water navy can undertake any of the above functions, a coastal navy is limited to being able to do inshore coastal tasks almost exclusively.

It is important to understand that the distinction between these two types of navies is one of purpose or national policy. It is not based upon ship size or type. It is based upon underlying naval theory, which guides doctrine, training, and strategy. Canada's navy, for instance, has over a century of experience and tradition firmly rooted in the blue water school inherited from the Royal Navy (RN). The fact that it eschewed battleships and large fighting squadrons after the Second World

War does not make it a coastal navy or one that subscribes to *guerre de course*, or maritime running battle; the Canadian navy remains a *blue water* force, committed to the classic strategy of *guerre d'escadre*, or fleet engagement – albeit with reduced possibilities for employment due to fleet size.

Clearly, this binary view of naval warfare is somewhat unsophisticated, but our investigation of naval theory is only a small component of our larger study. However, an examination of war at sea is a necessary complement to the study of war on land. Accordingly, opinions on naval warfare have been recorded for as long as mankind has gone down to the sea in ships. Thucydides stressed the importance that sea power, navies and control of the Ionian Sea had upon the early development of Hellas. Vegetius, too, talked of naval power. In the anonymous and unattributed 15th century English poem, *Libelle of Englyshe Polycye*, the need of the English to control the Straits of Dover to protect themselves from French invasion during the Hundred Years' War, as well as the benefits of sea power to the creation of colonies, was expounded; but none of these examples can be considered as true naval theory. Despite much writing about sea power, realistically, and practically, all of what was written prior to the 19th century was either a celebration of a naval victory or merely a collection of discourses on naval tactics.

It was not until the latter half of the 19th century that formal naval theory came of age, when historians and naval thinkers first undertook comprehensive critical analyses of maritime strategic doctrine. This work launched an era that became a golden age. In France the so-called *Jeune École*, (to distinguish it from the *Ancien École* thinking that it was attempting to supersede) under the guidance of Admiral Hyacinthe Aube, was questioning French naval policy with the implicit promise of an inexpensive counter to the cost of building and maintaining large blue water fleets. In Great Britain, Julian Corbett, Captain of Marines John H. Colomb and his brother Vice-Admiral Phillip Colomb were writing and lecturing on the influence of sea power upon historical development as well as the success and power of the British Empire. In the United States, Captain A.T. Mahan studied the influence of naval strength upon nations throughout history and lectured on this topic at the newly formed US Naval War College. In Germany, *Großadmiral* Alfred von Tirpitz was espousing unrestricted submarine warfare while building his High Seas Fleet. Like no other era before or since, the close of the 19th and dawn of the 20th centuries saw naval theory flourish.

The period was one of more than just intellectual ferment, it was also one of technological advancement and doctrinal turmoil. Although we could characterise the era in any of these terms it can best be seen as one long continuous naval arms race where technological innovation was viewed as the determining factor for command of the waves – sail and wood giving way to steam, the internal combustion engine, and steel. Incongruously, although the period from the end of the American Civil War until the dawn of the First World War was a golden age of naval thought, it simultaneously became a period during which almost all admiralties ignored much of what the new theories offered and "naval thought lagged behind naval technology."[3] Ministers and their admirals turned instead to technology with the result that the age was ruled by a form of technological determinism. Since admiralties and politicians turned away from the theorists, the real arguments on naval theory and their impact upon naval strategy occurred outside the naval ministries.

The fora for the exchange of great thoughts on naval warfare came in the private sector. Naval institutes and associations sprang up on both sides of the Atlantic during this time. In Britain, the Royal United Services Institute published essays not just on things military but also naval articles. In France, naval ideas were developed and debated in the *Revue Maritime*; in Russia *Sbornik* gave them voice. Italy's *Rivista Marittima* and Germany's *Marine Rundschau* did the same in their respective countries. The United States Naval Institute was founded in 1873 and began publishing its *Proceedings* that same year. Some naval annuals began during the period, such as those published by Thomas Brassey (*Brassey's Naval Annual*) and Frederick Jane (*Jane's Fighting Ships*), the latter of which continues into the present.

The fascination with emerging maritime technology manifested itself in several realms: analysis of the various small engagements that occurred was conducted almost exclusively from the aspect of equipment, whether it pertained to propulsion, gunnery, or armour. In this way, technology maintained the greater influence on both naval strategy and naval policy. Naval theory took a back seat to naval architecture and marine engineering; naval warfare attempted to re-invent itself through better engineering.

Ironically the era saw no protracted naval campaigns, no grand decisive battles, in which to test the competing theories. This may have been one reason why the limited engagements and skirmishes, which did occur, had such wide influence. For instance, the Russo-Turkish War of 1877 and the War of the Pacific of 1879, combined with advances in gunnery techniques and bore size, as well as the armouring of battleships, caused admiralties to look away from naval theory and to focus upon how technological developments were influencing tactics at sea. Nonetheless it remains an unanswered question as to what large protracted naval engagements might have wrought – had there been any; "the most likely outcome of an ocean campaign would have been indecisive and spasmodic encounters, and an ineffective use of seapower. It is ironic that during this period, the dogma of the decisive sea battle emerged as the centre of seapower theory."[4]

Theorists and Strategists

La Jeune École (1870-1905)

As a precursor to the coming golden age, there arose in France a group of naval officers that proposed countering British naval dominance by adopting a strategy of expanded *guerre de course*. The group caused a schism in French naval thinking, separating the navy into the 'old school' and the 'young school.' The latter, the *Jeune École*, soon took over the navy. Although an intellectual home for many young officers, the *Jeune École* also counted among its members the Minister of the Marine, several senior admirals, as well as prominent politicians and journalists. After the loss at Trafalgar in 1805 and France's ultimate defeat in 1815, the French navy had accepted that a continental power like France could not afford the resources required to challenge Great Britain's dominance of the seas. Thus, although the navy had been influenced by the results of the American Civil War and the Franco-Prussian War, the group's insistence that the era of blue water warfare had passed can be traced back to its philosophical roots after the Restoration of 1815.

One case captured the imagination of the *École*: that of the Confederate blockade-runner CSS *Alabama*. "The *Alabama* and other raiders did not succeed in breaking the Union blockade of the South or in significantly interrupting the sea communications of the North, but they did have enormous influence on naval thinkers of the time."[5]

The French loss to Prussia in 1871 spurred the movement. The Franco-Prussian War demonstrated that although France had had a large fleet, it had been made almost irrelevant by circumstance. The French fleet not only outnumbered the Prussian fleet, but it was also better. Nonetheless, Bismarck and Moltke, striving for a short war, kept their own ships in harbour, thereby denying the French the ability to use their naval superiority. The French took the sailors from the ships and used them to defend Paris, thus creating the odd situation where the navy, although not able to defend the republic, *per se*, became the heroes of a lost war. Philosophically, in military matters, the French went over to the defensive in all forms. This put a considerable strain on the French treasury. The new technologies of steam locomotion and torpedoes combined with this financial burden to breathe new life into the almost dormant belief in an inexpensive way for France to protect herself from British naval hegemony. Not for the first time, new technology was heralded as a saving strategy; the torpedo boat could be used to implement a strategy of *guerre de course*, simultaneously strangling British trade and protecting the French coast. It was a serendipitous alignment of necessities, the need for defence and financial constraint:

> ...coincided with the advent of one of the most persuasive and alluring theories of naval warfare, one which appealed to naval theorists and politicians alike throughout Europe. The Jeune École combined coastal defence, anti-blockade and commerce destruction into a package that promised a formidable naval presence at a reasonable cost. The admiralties of Europe were forced to take into account the threat a Jeune École-style

navy posed to modern ironclads, close blockade, amphibious assaults and maritime commerce.⁶

The movement was led by Aube who, after the Franco-Prussian War, had begun expounding his theories. In 1886, he became the *Ministre de la Marine*, and the movement gained not only support among sailors but also among such influential politicians and journalists as Georges Clemenceau, and Charles Pelletan. Aube had created a complex theory that combined the use of cruisers, fast gunboats, the *bateaux-canons*, and torpedo boats as part of his *guerre de course* strategy. Unfortunately, his tenure as minister lasted only eighteen months, having been in power long enough to arrest the construction schedule of French battleships, thereby giving impetus to cruisers, destroyers, and torpedo boats, but not long enough for his complex theory to be put into practice or even to be fully understood.

In the end, the *Jeune École* did not serve France well and deservedly slipped into historical oblivion. "Never had France spent so much money for its navy with such deplorable results."⁷ Their claims of technological solutions to France's geostrategic problems fell well short of the mark: "Aube mistakenly felt that technology made moral factors irrelevant; many of his ideas, such as about torpedo boats, did not work; and he and his followers tried fruitlessly to make up for their mistakes with complex mathematical models. The result ... was decades of disarray for the French Navy."⁸

John Colomb (1838-1909) and Phillip Colomb (1831-1899)

Although now almost forgotten, the Colomb brothers were seminal in the use of history to assist in the advancement of naval theory. Specifically, they pioneered historical analysis as it related to British imperial defence policy. Significantly, they preceded both Mahan and Corbett. Captain Sir John Colomb, the younger brother first declared Great Britain's need for a global network of naval bases in 1867, and Vice Admiral Phillip H. Colomb expanded on the same issue in the 1880s and in 1891.⁹

> England alone possesses the appliances for making this transfer in every quarter of the globe. Her enemies must commonly load with coal in neutral ports, in short measure, and in haste and fear. ... [The Empire] is a vast, struggling nervous, arterial, and venous system, having its heart, lungs, and brain in the British Islands, its alimentary bases in the great possessions of India, Australia, and North America, and its ganglia in the Crown Colonies. ... Main arteries and corresponding veins lead east through the Mediterranean and the Red Sea to India, China, and Australia; west to America and the West Indies; south to Australia, Southern Africa and

America, and to the Pacific.[10]

John Colomb was a Royal Marine artillery officer. His older brother retired from the active list as an RN Captain and was promoted on the retired list. Their real contributions were primarily in naval tactics and thereafter in strategy, with Phillip writing and lecturing extensively on the issue sea power and how it could be used to sustain and promote imperial defence. Although many of his articles on the subject preceded the publication of Mahan's book, the American soon eclipsed both him and his brother. Not long after Mahan, Corbett relegated both brothers to obscurity and robbed them of any real influence.

Alfred Thayer Mahan (1840-1914)

In the English-speaking world, Mahan is usually the first name that surfaces when discussions turn to naval warfare theory. "No other single person has so directly and profoundly influenced the theory of sea power and naval strategy..."[11] And after the publication of his book, *The Influence of Sea Power on History, 1660-1783*, no one held such sway in the world naval community. The son of a West Point professor and highly respected army officer, this successful naval officer leapt to world fame in 1890 with the publication of his work. The book was based upon his US Naval War College lectures, which stemmed from his long study of world history.

Mahan's book had, in fact, been at least partially published as a refutation of the naval theories proposed by the *Jeune École*.[12] Whatever influence the French group may have had in their own country, arguably their greatest influence was because of the American taking exception to their ideas:

> The *Jeune École* may have had its greatest effect, however, through its influence on a naval strategist who disagreed strongly with most of its program, Alfred Thayer Mahan. Mahan wrote his major works at least in part to counter its influence, seeking specifically to disprove the view of the Jeune École (and many others at the time) that the days of great naval battles were past.[13]

What made Mahan's text so exciting was that it was the first book to formulate broad principles of war for navies. Prior to his success, books on naval history and warfare had focused almost exclusively on tales of heroic battles, engagements at sea and memoires. Even a casual perusal of any of the tomes written before Mahan bears out this point. Naval history was either one step removed from the heroic fiction, so popular in the 20th century in the guise of C.S. Forester's Horatio Hornblower, or it was the stuff of diagrams, tactical deployments, who had the wind or lee and a listing of names of ships and their captains. Mahan avoided this temptation, instead being the first to formulate naval principles of war in universal terms, choosing a different tack from those who had preceded him. He approached sea power as Jomini had done on land. But to consider Mahan as the naval Jomini is perhaps unfair for however much he may have been influenced by his Swiss predecessor, and there is no doubt that he was, he was not quite as dogmatic as the comparison would imply.

Mahan's thesis was that sea power had been the determining factor in establishing the great empires of the past. He looked at Rome, Spain, Portugal, and England and how their navies had been used to build their empires. Rome's ability to attack Carthage with impunity, for instance, was a direct consequence of its powerful navy. Even more recently, the Napoleonic Wars demonstrated that sea power and not just the armies of the Grand Alliance had sealed France's fate.

Mahan believed that Britain's mastery of the sea was a major and perhaps misunderstood contributor to the defeat of Napoleon. After Admiral Horatio Nelson's naval victory at Trafalgar, on 21 October 1805, it took ten years, but eventually British sea power played a decisive role in bringing the Napoleonic Empire down. Trafalgar all but destroyed the combined French and Spanish fleets with no ships-of-the-line lost by the British. But the importance of this victory lay not in the loss exchange ratio; it lay in confirming Britain's undisputed command of the oceans. Britain could then concentrate on fighting Napoleon on the continent while simultaneously impeding the movement of supplies to France and her allies as well as obstructing troop movements by sea. Mahan argued that the British naval blockade of France after Trafalgar ultimately forced Napoleon into his policy of 'continentalism' and led to his disastrous march into Russia from which he, his armies, and France never recovered.

At its heart, Mahan's theory had a dual theme: first, that the essence of sea power was the ability of a nation to use the oceans to go wherever and whenever it wished – and that this freedom of action was to be denied to the enemy; and second, and perhaps even more important, that the primary purpose of fleets was to engage other fleets in decisive battle, what was known as *guerre d'escadre* or squadron warfare. In other words, battle fleets needed to seek out and destroy enemy battle fleets:

> Mahan believed that, first and foremost, the true objective in a naval war was always the enemy fleet. The traditional weapon of the weaker naval power, commerce raiding (*guerre de course*), would not be decisive without support of squadron warfare. Battle was to be carried through to a decisive finish, naval force was to be used offensively, and the fleet had to be concentrated and never divided. He considered capital ships (in Mahan's time, battleships) the most important vessels in the navy and found blockade valuable, a central position and interior lines advantageous, and proper overseas bases important for naval operations. According to Mahan, one obtained "command of the sea" by concentrating one's naval forces at

> **MAHAN's FACTORS**
>
> 1. Geographical Position
> 2. Physical conformation
> 3. Extent of its territory
> 4. Size of its population
> 5. National character
> 6. Character and policy of its government

the decisive point to destroy or master the enemy's battle fleet; blockade of enemy ports and disruption of the enemy's maritime communications would follow.[14]

His study of history convinced Mahan that there were six factors that strongly influenced the ability of a nation to gain sea power: a state's geographical position; its physical conformation; the extent of its territory; the size of its population; national character; and the character and policy of its governments. Although Mahan's assertions now seem almost self-evident, when they were made, they caught the imagination of anyone interested in the use of sea power. Like Darwin before him, he opened the eyes of all those who followed. Also, like Darwin, his writings became highly distilled, oversimplified, and interpreted. In Mahan's case, all naval warfare was reduced to the notion of great decisive battles at sea. Clearly influenced by Jomini and associated with him probably because of his father's influence, both Jomini and Clausewitz are easily discernable in his work even though Clausewitz said little or nothing about naval warfare save for some small comments on amphibious operations.

Mahan was the voice of naval theory for his age. Almost every navy in the world studied his writings, and he was feted wherever he lectured. Through the persuasive tool of historical analysis, he convinced not only the America but also most other industrialised nations that command of the seas held the key to national greatness. His influence has been both pervasive and long-lived, with the world's great blue water navies, the Royal Navy, and the US Navy, still teaching his ideas. Even The People's Republic of China, with its push to build capital ships and aircraft carriers has come under his sway. Nonetheless, Mahan's assertions were perhaps too rigidly embedded in the past; his studies were based on history, with the result that he tended to depreciate issues like technology, trade protection, amphibious operations, blockade, economic power, and limited warfare.[15] So, although still strongly influential and required reading for naval officers the world over, Mahan's arguments continue to engender debate.

Julian Corbett (1854-1922)

Mahan's undisputed ascendancy did not last long. Two decades after the American sailor rose to world pre-eminence, a British historian turned his lectures at the Royal Naval College in Greenwich into a book to rival *The Influence of Sea Power*. In 1911, with an obvious nod to Clausewitz, Corbett published *Some Principles of Maritime Strategy*. Corbett had drawn his conclusions from his own studies of history, in particular Britain's strategy in the Mediterranean during the Seven Years' War.

He claimed that his investigation had disclosed that a fuller understanding of sea power would better allow its political and tactical application and produce results out of proportion to the actual forces employed. Corbett did not completely disagree with Mahan, but he did feel that the American held a too rigid interpretation of what command of the sea could mean. He denied Mahan's insistence that battle was the goal of naval fleets. In opposition to Mahan, Corbett underscored the connection of naval and land warfare. Further, he concentrated on the importance of lines of communications rather than battle fleets engaging one another. Corbett was quite clear that the purpose of naval strategy was a question of communications: "The power of the second method, by controlling communications, is out of all proportion to that of the first – direct attack. Indeed, the first can seldom be performed with any serious effect without the second."[16]

Corbett argued that since communications ware an essential prerequisite of all maritime operations, whether naval or commercial, ensuring the security of sea lines of communications (while denying the enemy the same) was an important goal in and of itself. Fleets were therefore to be used primarily to secure communications, not to seek out enemy fleets for battle. In other words, Corbett posited as a strategy the reverse of what Mahan had proposed. Further, Corbett believed that control of the sea allowed the stronger nation to use the tactics of raids (*guerre de course*) to isolate strategic objectives on an enemy's territory while protecting one's own strategic objectives. He believed that the RN had created a sea power strategy that focused on its exertion of economic pressure ashore through blockade and embargo with only limited engagements on the littoral. This "raiding activity would be valuable against an enemy whose army was too strong to be attacked on the battlefield, and such raids should therefore be part of any strategic plan."[17]

Corbett's views were quite controversial since his downplaying of the importance of the decisive naval battle was seen in navy circles as a denigration of the RN's cherished Nelsonian tradition. And his personality did nothing to ameliorate the situation. The controversy began with Corbett's refusal to accept Mahan's definition

WAR AT SEA

of command of the sea. Instead of seeing the definition in terms of one fleet having won a crushing victory and having swept the seas of opponents, Corbett had a more nuanced interpretation of what sea power meant (or could mean). He believed that there could be different kinds of maritime hegemony, based upon national desires and objectives. Rather than work purely from the premise that the enemy's fleets must first be destroyed, Corbett insisted that the security of maritime communications could be achieved in various ways based upon expediency as well as national aims. His views held great influence in England but were criticized by men like Spenser Wilkinson because they denied Mahan's primacy of battle.

Corbett's thesis was viewed as heresy by many in the RN to the point where he was blamed for the navy's poor showing at the Battle of Jutland.

> [His] original ideas embroiled Corbett in a vitriolic debate with some of the leading military theorists and naval experts of his time. This most assuredly did nothing to enhance his reputation since the subtlety of his ideas destined them to be misunderstood. For example, Lord Sydenham later accused him of exerting a negative influence on the doctrine, plans, and morale of the British navy, thereby contributing to their failure to achieve decisive results in the battle of Jutland. Many years later, Cyril Falls charged Corbett with "minimizing the importance of combat." Despite a barrage of criticism, Corbett steadfastly refused to change his strategically "blasphemous" conclusions.[18]

In modern parlance, Corbett believed that naval power could act as a combat multiplier, allowing one side to gain disproportionate advantage whether at the tactical, operational, or strategic level. Corbett chose an example from the Napoleonic Wars: The British, because of their secure lines of communications, could menace the French almost anywhere along their coast with the threat of amphibious assault. This threat "was always out of all proportion to the intrinsic strength employed or the positive results it could give .... Its value lay in its power of containing force greater than its own."[19] Corbett quoted Napoleon, citing an example where England's naval superiority gave it a tenfold advantage over the French: "With 30,000 men in transports at the Downs the English can paralyse 300,000 of my army, and that will reduce us to the rank of a second-class Power."[20] However, due to the many years it takes to build warships, Corbett's book arrived too late to alter the pre-war naval arms races that Mahan's theories had spurred. This was unfortunate, for his work was both broader in scope and more profound in its understanding of the complexities of the modern age. In many ways, Corbett completed the work begun by Clausewitz. In fact, Corbett did more than just fill the gap that Clausewitz had left; "In the process of brilliantly adapting Clausewitz's theory to the unique circumstances of naval warfare, particularly to the needs of

British strategy, Corbett actually developed his own innovative theory of limited war in maritime strategy."[21]

Herbert Richmond (1871-1946)

Admiral Sir Herbert Richmond was a highly respected seagoing officer who was not only a student and supporter of Corbett but, also, the leader of the 'young Turks' in the RN during his naval career (1885-1931). He actively agitated for reform throughout his career and, after retirement, lectured at Oxford on sea power. His lectures were later compiled as *Statesmen and Sea Power*, published in 1943. He is perhaps best remembered for his statement: "Command of the sea is the indispensable basis of security, and whether the instrument which exercises that command swims, floats or flies is a mere matter of detail."[22] After the First World War, Richmond was among those who were condemnatory of the RN's contribution to Allied victory. Like the Colomb brothers, Mahan, and Corbett, Richmond had conducted his own historical studies of naval warfare. Convinced of the importance of defending trade and of the use of convoys as a means doing so, he downplayed the central importance of the Mahanian 'decisive battle' in favour of the more flexible approach taken by Corbett.

The World Wars

Theory is only useful if it can be applied. Not long after Mahan and Corbett wrote their theses, the world went to war, and many saw the opportunity to convert theory into practice. Unfortunately, the test proved inconclusive. The first trial of the theoretical conflict between Mahan and Corbett came at the Battle of Jutland at the end of May 1916, where the German High Seas Fleet met the RN's Grand Fleet in what was to be the only major naval engagement of the war. The battle was a classic naval battle with long-range gunnery, dreadnoughts, and battle cruisers all playing their respective roles. Over the course of approximately twelve continuous hours of firing and manoeuvring, the two commanding admirals, Sir John Jellicoe, and Reinhard Scheer, parried and thrust with their forces. Both sides claimed victory – the Germans, a tactical one for they had sunk more tonnage than the British; the British, an operational one for the High Seas Fleet never again ventured forth to challenge the Royal Navy's control of the seas. Regardless, one issue was clear: "The duel at Jutland had demonstrated the futility of material strategy."[23] Technology had trumped tactics:

> [I]n the new seamanship of iron and steam, mathematics were subverting the art of centuries, and vistas of possibilities opened up for tightly choreographed geometrical evolutions. The 'science' of Steam Tactics was

the result, and every movement, every change of course speed or formation, could be ordered and executed by flag signal ... [The] goosestepping doctrine was consonant with Victorian notions of order and propriety.[24]

Much like warfare on land, that at sea did not conform to pre-war predictions. The climactic engagements and battles upon which Mahan had based his theories did not occur. Sufficient RN commanders had fallen under the sway of Corbett and so the RN had not sought such engagements. However, there was almost no consideration given to the possibility that Britain could be economically vulnerable to a concerted effort by Germany to disrupt her shipping. The full consideration of what sea power provided to the Empire's economic lifeblood "was somehow not preserved in the progression through the naval race of the first decade of the 20th century."[25] The RN gave little credence to the threat. Meanwhile, the German U-Boat fleet pursued a strategy of *guerre de course* and very nearly succeeded. It was more good luck than an effective anti-submarine strategy that saved Britain during the First World War:

> In the end of course, both Corbett and [Admiral Lord] Fisher were proved right: The enormous naval and mercantile superiority of Great Britain proved simply too much for Wilhelmian (sic) Germany to absorb. The point however, is surely that the complacent British attitudes and mediocre analytical abilities that were prevalent at the time, were ill deserving of this good fortune.[26]

The Second World War once more raised the question of whether Mahanian or Corbettian theories were more valid. Mahan's major fleet-on-fleet engagements, which would destroy an enemy navy and grant command of the sea, did not occur. Although major engagements did take place – the Battle of the Atlantic, the Coral Sea, Midway and Leyte Gulf, for instance – they did not give command of the sea to either side; and, moreover, the greater use of submarines and naval aircraft complicated matters. "The more nuanced approach of Corbett as well as the latter's emphasis on the importance of combined operations seemed closer to the mark. On the other hand, Mahan's rejection of the *guerre de course* was vindicated by the Allied defeat of the German U-boat offensive...."[27]

Post-Second World War writing on naval strategy has been rich and varied, but nothing of real significance has emerged to rival the influence of either Mahan or Corbett. Whether we look to the 'all nuclear navy' proposed by the influential American Admiral Hyman G. Rickover (1900-1986) or the incredible (and eventually unsustainable) growth and expansion of the Soviet navy under Admiral Sergei Gorshkov (1910-1988), its Commander-in-Chief from 1956 to 1985, or the rapid growth in the 21st century to the Chinese People's Liberation Army Navy, naval strategy has received growing attention. But this attention has been inextricably entwined with air and land power theories, and there has been a pressing concurrent need to understand warfare as a *joint* endeavour. Nevertheless, no one has yet emerged in the realm of naval theory to displace either Mahan or Corbett. The golden age of naval theory remains firmly ensconced in the past.

The Submarine and Naval Aviation

The development of air power altered the employment of navies and heralded a host of changes, including the emergence of aircraft carriers, naval air fleets and the development of submarine-based intercontinental missile forces. What it did not alter, however, was the fundamentals of naval theory. The creation of ships that could carry aircraft was merely a way of extending the striking power of fleets. Thus, aircraft did not so much cause a revolution in military affairs as offer extended possibilities for admirals to project naval power without hulls in the water.

Sea-based aircraft during the First World War are barely worth discussing, but the use of both land-based and sea-based aircraft during the Second World War showed that command of the seas could be greatly influenced by whoever controlled the air space above those seas. The Pacific Campaign of the Second World War saw the most extensive use of naval air forces in battle. Due to the vast distances faced by the theatre commanders and as a counter to the opponent's use of air power, great advances were made in the use of aircraft to augment naval power in terms of reconnaissance, offensive striking capability, and defensive screening of fleets. In terms of sea power, aircraft demonstrated a great ability to destroy enemy shipping but were almost completely useless to gain sea control. Nonetheless, none of these uses altered the basic premises of naval power from either a Mahanian or Corbettian perspective.

Although first introduced during the American Civil War, the submarine only came into its own as an independent fleet arm during the First World War. On all sides, in both World Wars, submarines had a profound impact upon naval strategy and tactics. In both struggles, the submarine was employed almost exclusively as a commerce destroyer as a *guerre de course* strategy. The submarine's great strength – stealth – was also its greatest weakness. As with aircraft, submarine fleets like the Germans built in the early 1940s, and the Soviets built in the 1970s, could deny shipping safe passage but could not by themselves gain command of the sea.

Yet, a major development did come in the 1960s. The use of long-range guided missiles launched from nuclear-powered submarines transformed the submarine from an auxiliary arm subordinate to surface actions into a major weapon of strategic bombardment. The new nuclear-powered boats carrying guided missiles proved to be virtually invulnerable to detection and attack. The Soviet navy took full advantage of this stealth capability to passively threaten the American homeland from thousands of miles offshore. In the Soviet view, technical advancements in submarines made them weapons systems that could be use at the strategic, operational, and tactical levels all at the same time.[28] The Americans, too, availed themselves of this flexibility, demonstrating just how effective a weapon these strike submarines could be during the First Gulf War. In that operation, submarine-launched Tomahawk missiles struck Baghdad from hidden locations in the Indian Ocean and the Eastern Mediterranean with terrifying accuracy. But let us be clear: submarines can *deny* sea lanes of communications to an enemy, but they *cannot* command or control them.

Summary and Conclusions

Despite being almost as old as warfare on land, warfare at sea suffered from a much later start in the development of its own theories. By the time naval theory reached its classic age (and arguably its high-water mark), theorists of land warfare had already made their marks and had thereby laid the foundations of modern naval theory. In the words of Michael Handel:

> [W]hile Mahan integrates and synthesizes Jomini's work with his own, Corbett uses Clausewitz's *On War* as a heuristic point of departure. Mahan, in other words, remains loyal to Jomini's ideas, and by extension, those of the 'continental strategists.' In contrast, Corbett, although inspired by *On War*, develops ideas different from and sometimes contradictory to those of Clausewitz. The subtle approach adopted by Corbett ironically resembles that of a work he had never read – Sun Tzu's *The Art of War*.[29]

Notwithstanding the late start, naval theory, whether seen as standing alone or as an outgrowth of classical land-based theories, quickly became a fully formed and necessary independent component of Military Theory. The fact that it has not been built upon two millennia of discussion, as with warfare on land, does not make naval theory any less important.

Navies have political utility that armies and air forces have always lacked. Warships and fleets have peacetime political uses that are, arguably, almost as important as those of wartime. Navies can project power; they can be instruments of foreign policy; they can promote trade. "When Theodore Roosevelt wished to announce that America had come of imperial age, he sent neither ambassadors nor armies around the world but the Great White Fleet."[30] If for no other reason, the political importance of navies as tools of statecraft make the study of naval theory of primary importance. Ignorance of the subject – or a misinterpretation of events – can lead to errors with long-term consequences.

The development of naval warfare offers the student of Military Theory an excellent demonstration of how technology may appear to change the nature of war, when what is changing is its character. Over the course of history, human power was replaced by wind power; the use of sail was replaced by steam created by coal, then oil-powered engines, and then nuclear power. Setting aside the tactical implications of such technological advancements, the determining factor regulating the operational and strategic use of sea power has swung like a pendulum. Originally, human endurance was the determining factor controlling how long fleets could remain at sea. The industrial age then made fuel sources the key to fleet endurance, and then nuclear power once more re-established the primacy of human endurance. The ability of ancient fleets to dominate sea-lanes was restricted primarily by the ability to harness human muscle power. Whether we talk of manning triremes or men-of-war, this restriction changed little from Roman times to the age of sail. The primary reason that ships put into harbour was to relieve crews and to revictual. However, even this reason was not absolute, as the Royal Navy developed a system of resupplying its ships so that they might remain on station for weeks or even months. The introduction of coal-fired boilers and steam propulsion predicated the need for coaling stations strategically located around the globe. Coaling stations soon gave way to oil fuelling stations; but the delimiting factor remained the needs of the ship and not the crew. With the advent of nuclear propulsion, the restriction returned to the human side of the equation. The delimiting factor determining the endurance of fleets at sea was not the wind or the amount of coal aboard or the size of the fuel tanks. It had nothing to do with the technical aspects of the ship. The critical factor determining fleets' deployment was the ability of crews to be sustained at sea. Ironically, what this meant was that although warships gained more tactical mobility, fleets had less strategic mobility than in earlier times. Focusing only upon the technological advancements could cause admirals and politicians to forget the human restrictions that these technologies re-imposed. Failure to appreciate this restriction could result in the inappropriate use of fleets – both in war and in peace.

Whatever else we may deduce from the study of sea power theory, one major lesson is clear: the cornerstone of all sea power theory is the need for command of the sea. The real question is: What is meant by command of the sea? The question is more complex than it may appear at first for, in essence, it is this definition and the resultant implications thereof that separate Mahan and Corbett. Mahan's greatness lay not only in being the first to recognize and demonstrate the importance of naval power upon national destinies but also in postulating that command of the sea was the real object of all naval power. Just as important, this object was to be gained by the offensive action of a blue water fleet seeking to achieve a concentration of force against an opponent's fleet in a *guerre d'escadre*. Mahan felt that states needed fleets 'in being' because the risk that a state could be blocked from building or purchasing a fleet would be too great should it lose control of its sea-lanes or have its commerce blockaded. Mahan rejected the notion of *guerre de course* as a primary object of any fleet; he felt that it was ineffective. Once an enemy fleet was destroyed, the enemy could easily be blockaded or attacked, thus making a strategy of *guerre de course* unnecessary.

Corbett disagreed with Mahan. He felt that his definition of command of the sea was too narrow. Considering technological change, temporary control for the conduct of specific operations was more realistic. This obviated the need for a decisive battle. He also did not believe that the ocean was the decisive theatre of war and so a navy's real contribution was in how it assisted land armies defeat an enemy. Unlike Mahan, he accepted that navies could fight a series of small engagements, that command of the sea could be measured in degrees. In other words, that a strategy of *guerre de course* combined with amphibious operations was not only an acceptable strategy for a navy, but it was also preferable. Decisive battles were too risky for even if a navy won such a battle, it might be seriously weakened and therefore command of the sea could be put at risk by a third party. It was exactly this corollary that the German High Seas Fleet developed in its bid to challenge the Royal Navy's mastery of the seas. Tirpitz's so called *Risikogedanke* threatened to damage the British fleet such that, although perhaps not fully beaten, it would no longer have the capacity to dominate the world's oceans. It was this threat that kept the British fleet bottled up at Scapa Flow during the First World War. Influenced by Corbett, Jellicoe was hesitant to bring the fleet out for a major action against the Germans and, when he did finally engage them, the result was unsatisfying to both navies.

Thus, although the two great naval theorists agreed upon the underlying fundamental premise upon which naval power should be built, their interpretations, and their conclusions, took them in radically different directions. Although both theories were predicated upon the concept of command of the sea, the resultant theories used this term to mean different things. The two schools that grew up based upon these theories, therefore, had sometimes radically different views of the purpose, structure, and employment of navies.

Finally, the history of the French *Jeune École* is an object lesson in intellectual narcissism. Realizing that it could never topple the Royal Navy from its position of blue water dominance, the French Navy became enamoured of a tantalizing mixture of technology and *guerre de course*. The idea of technology making up for physical shortfall may have been an interesting intellectual exercise, but the reality of the situation fell far short of the promise. The history of Military Theory is replete with such instances, where attractive ideas that were full of promise led to a serious degradation of military effectiveness. A good idea, but an untested theory, is not enough; it must be both applicable and proven to be of any use. In the case of the French Navy, its generation-long flirtation with the *Jeune École* degraded its usefulness as an instrument of national policy and cast it into a disorder from which it only recovered after the Second World War. The history of the *Jeune École* should be an object lesson to all on the importance of studying – and understanding – Military Theory.

STRATEGIA

Notes

1. Edward N. Luttwak, *The Political Uses of Sea Power*, (Baltimore, MD: Johns Hopkins University Press), 1-2.
2. Herbert W. Richmond, *National Defence: The Navy*, (London: William Hodge & Co, 1937), 30.
3. Margaret Tuttle Sprout, "Mahan: Evangelist of Sea Power," Makers of Modern Strategy: from Machiavelli to the Nuclear Age, (Princeton, NJ: Princeton University Press, 1986), 415.
4. J.R. Hill, "Accelerator and Brake: The Impact of Technology on Naval operations, 1855-1905," Journal for Maritime Research, (December, 1999).
5. Eric J. Dahl, "Net-Centric Before its Time: The Jeune École and its Lessons for Today," Naval War College Review, (Autumn 2005), 113.
6. David Harold Olivier, "Staatskaperei: The German Navy and Commerce Warfare. 1856-1888," unpublished PhD diss., (University of Saskatchewan, 2001), 207.
7. *Jamais la France n'avait dépensé autant d'argent pour sa marine avec des résultats aussi déplorables.* "Le Technicisme et la Jeune École", Institut de Stratégie Comparée (ISC), École pratique des Hautes Études, Sciences historiques et philologiques, Sorbonne Paris, available at http://www.stratisc.org/PN4_MOTTEMAHAN_3.html. (My translation.)
8. Dahl, 111.
9. Clark G. Reynolds, *Command of the Sea: The History and Strategy of Maritime Empires*, (New York, NY: William Morrow & Co., 1974), 414.
10. Philip H. Colomb, *Essays on Naval Defence*, (London: W. H. Allen and Co., 1893), 35-7.
11. Sprout, "Mahan", 415.
12. Theodore Ropp, "Continental Naval Theories," Theodore Ropp, "Continental Naval Theories," Makers of Modern Strategy from Machiavelli to Hitler, Edward Mead Earle, ed., (Princeton, NJ: Princeton University Press, 1973), 446.
13. Dahl, 122.
14. Ibid.
15. Reynolds, Command of the Sea, 415.
16. Julian S. Corbett, *Some Principles of Maritime Strategy*, (London: Longman, 1911), 337.
17. Paul G. Halpern, "Theorists of Naval Warfare" available from http://college.hmco.com/history/readerscomp/mil/html/mh_0366_navalwarfare.htm. Note: this is an archived page and must be accessed using an Internet Archive service.
18. Handel, Michael I. (2000) "Corbett, Clausewitz, and Sun Tzu," Naval War College Review: Vol. 53 (2000): No. 4, 109.
19. Corbett, 67.
20. *Correspondance de Napoléon*, xix, 421, 4 September. As quoted by Corbett, 69.
21. Handel, "Corbett and Clausewitz", 117.
22. Herbert Richmond, *Statesmen and Sea Power* (Oxford: Clarendon Press, 1946), 136.
23. Reynolds, 463.

24. Gordon Andrew, *The Rules of the Game* as quoted in Frank Snyder, "A Matter of Interpretation," Naval War College Review, (Spring 1998).
25. Angus Ross, "Losing the Initiative in Mercantile Warfare: Great Britain's Surprising Failure to Anticipate Maritime Challenges to Her Global Trading Network in the First World War" International Journal of Naval History, (April 2002), 2.
26. Ibid, 10.
27. Halpern.
28. Sergei G. Gorshkov, *The Sea Power of the State*, (Willowdale, ON: Pergamon Press, 1980), 224.
29. Handel, "Corbett, Clausewitz, and Sun Tzu," 107.
30. Michael Kelly, "The Air-Power Revolution", Atlantic Monthly, (April 2002), 21.

> "To conquer the command of the air means victory; to be beaten in the air means defeat and acceptance of whatever terms the enemy may be pleased to impose.

General Giulio Douhet
Command of the Air

WAR IN THE AIR

Introduction

We now turn our gaze skyward to the realm of air power theorists and strategists. Based upon one or other philosophies, the men discussed below created theories that allowed humankind to better understand air power both conceptually and as an instrument of war. Our focus will be primarily upon the theorists Giulio Douhet and William Mitchell. Other theories and strategies such as those proposed by colonels Boyd and Warden are only discussed to round out the picture and we will revisit them later in another chapter. Also, we will discuss the shifts of military thought and theory during the Cold War since this development was predicated primarily upon air power theory. The chapter's value is in gaining an appreciation of both the influence of land theorists upon the development of war in the air as well as the influence that air power theorists have had upon gaining an insight into the nature of war through a better understanding of war-centred air power theory.

More so than war on land or at sea, the conduct of war in the air has been shaped by technology. The rapid development of technology has frequently led some apostles of air power to mistakenly claim that the advent of the airplane forever changed the nature of war when what they were actually seeing was a rapid change in its character. From the assertion by Douhet that air power could win future wars without the need of ground troops to the declaration by Mitchell that navies were obsolete, the claims ignited and re-ignited debate throughout Western society. As with war at sea, war in the air is fundamentally straightforward: it can be broken into two broad generic sub-groups: fighters and bombers. Clearly, such a crude classification is a gross simplification. Air warfare includes the aspects of unmanned vehicles, rotary wing flight, strategic, operational, and tactical levels, air-to-air combat, sub-orbital missiles, naval aviation, and many, many more. None-the-less, for the purposes of discussing air power theory, the bifurcation of air power into either 'fighting' or 'bombing' is a useful simplification.

Fighters destroy targets that are usually, but not exclusively, in the air, whereas bombers destroy targets on the ground or at sea. From an air power perspective, fighters are most closely, but not exclusively, associated with tactical air power while bombers are likewise associated with strategic air power. Once again, this is an overly simplistic view of a highly complex subject but one that offers the general student of war a reasonable point of departure from which to consider air power theory.

Little has captured the human imagination quite like the concept of having 'slipped the surly bonds of earth' as the nineteen-year-old poet and RCAF pilot John Magee famously wrote. From daydream to dangerous oddity to practical reality, war in the air has repeatedly held the promise of startling results at relatively low costs. In many cases, the prophesies had been bold: Prime Minister Stanley Baldwin's 1932 claim that 'the bomber will always get through' for instance or General Hap Arnold's 1946 assertion that the development of air power 'will reduce the requirement for, or employment of, mass armies and navies.' Decades of embarrassment over promises that air power has not been able to deliver has not curbed the near-hyperbolic enthusiasm of air power advocates: US Air Force colonel John A. Warden III, the father of the latest thinking on the use of strategic air power, went so far as to claim in 1997 that air power was the most effective method of fighting in human history.

Compared with the brutish bloodletting of combat on the ground, aerial warfare has consistently held out the promise of clean, almost 'antiseptic' killing, death from a distance. Like the promises of diehard sports fans for their beloved teams, the predictions made by the early proponents of air power were both broad and grand, frequently missing their marks:

> The history of air power has also been confused by the bragging of its prophets and the derision of its enemies. Too often vision has outrun reality and resulted in disappointment and reaction. As newcomers, forced to plead from a position of weakness, airmen carried their arguments to their logical extremes and talked about what air power was going to be able to do; and their listeners tended to forget that these were prognostications, accepting them instead as imminent realities.[1]

Initially, airplanes were not considered able to operate alone as an independent service but only as an ancillary to either ground or maritime forces; air forces did not snap the ties that tethered them to armies and navies until the Second World War. Nevertheless, to some of the early visionaries, the new air arm held the promise of becoming *the* deciding factor in war. Men like Douhet and Mitchell ignited debates on the efficacy of air power, of how it should be employed, as well as the perennial dispute regarding the proper balance between fighters and bombers, debates that continue to this day.

WAR IN THE AIR

The first significant progress in 'controlled' flight was made by a Prussian general, Ferdinand *Graf* von Zeppelin, who advanced the use of 'rigid' dirigibles for both military and civil purposes. Building his first dirigible in 1900, he captured world attention. It was not long before all dirigibles came to be known generally as *zeppelins*. The addition of engines and propellers to these airships gave them greater utility, but they were too large and slow. The need for greater speed necessitated the development of heavier than air machines and, in 1903, Orville and Wilbur Wright achieved the first 'controlled and sustained' heavier than air, powered, manned flight. Military interest was immediate but neither well-funded nor well-coordinated. It took until 1910 when an American experiment proved that these aircraft could become useful military weapons: a biplane was successfully launched from a warship (*USS Birmingham*). The following year Italy, at war with Libya, used various aircraft for aerial reconnaissance and, not long later, for dropping grenades on ground troops.

Manned mechanical flight and aerial warfare are really the progeny of the 20th century. The First World War saw the advent of military air power and with it the possibility that commanders might finally free themselves from the constraints of unfriendly terrain. Air power advocates like Douhet and Mitchell sought to demonstrate the utility of air power in general and especially of the bomber. They argued that air power offered a rapid and relatively inexpensive way to attack the enemy's vital points. The basic premise was straightforward: air power could quickly destroy an enemy's industrial war-making capability or even more directly attack the enemy's civilian population centres. Without the restrictions of geography, air power could strike directly at the enemy's vital centres of gravity, (recall Clausewitz) thereby changing the traditional arc of war. Douhet argued:

> A nation which once loses the command of the air and finds itself subjected to incessant aerial attacks aimed directly at its most vital centers and without the possibility of effective retaliation, this nation, whatever its surface forces may be able to do must arrive at the conviction that all is useless, that all hope is dead. This conviction spells defeat.[2]

Either way, through the loss of industrial capability or through the loss of the civilian population's will to fight, the enemy would be compelled to surrender. This was an enticing proposition. After millennia of battlefields littered with corpses, the idea that a war could be won by breaking the will of an enemy through bombing was certainly worth considering.

Theoretically, there are only three variables that affect weaponry: range, accuracy, and lethality. The first has been the primary driving influence; whether on land, at sea or in the air, the advantage has usually gone to the side that could see the furthest and then bring weaponry to bear upon an adversary. Armies would, therefore, seek high ground; navies would post lookouts at the top of their ships' masts. Although

STRATEGIA

the range of weapons kept increasing, the laws of physics limited the trade-offs among range, accuracy, and lethality. In the age of purely kinetic weapons, the farther it could shoot, the less accurate and less lethal it was.

The advent of airplanes suddenly increased range and lethality almost unimaginably. In some ways, all military theories have been chariots drawn by two horses: the first has been the constant search for technological advantage; the second has been the introduction of new ideas. The advent of air power suddenly held out the promise of practically limitless range coupled with heretofore unheard-of accuracy. Planes could reconnoitre the battlefield and bring back information that would allow a commander to manoeuvre his forces to maximize the range of his weapons or better yet, drop ordnance directly on an enemy from the sky. Even so, at least in the beginning, most commanders failed to grasp this promise fully, considering aircraft only in the context of providing more height than had previously taller ships and higher hills. As had often been the case, the technology was available in advance of the new idea. Technology had presented a new weapon. What was needed was someone to create a theory of how best to use it.

Theorists and Strategists

Giulio Douhet (1869-1930)

Few human beings can claim to have introduced a whole arena of military theory. Douhet was the first man to propose a theory based not only upon the use of a new weapon, but on the use of another dimension. The father of aerial bombardment was born in 1869 in Caserta, Italy. As a boy, Douhet entered the military academy as an artillery officer and was admitted to the war college in 1896. By 1915 he was commandant of Italy's first aviation battalion. He spread his ideas in writing and was critical of what he considered to be the narrow-minded and inflexible senior commanders in the army. Himself a stubborn man, he soon created enemies in high places.

Italy entered the First World War in May 1915 on the Allied side. Douhet was posted to the General Staff and not long after was inundating the Army Chief of Staff with memoranda regarding the massed used of air power. Ignored, Douhet increased his criticisms of the senior command. After writing a memorandum to the war ministry complaining of the ineptitude of the senior leadership, he was court-martialled and sentenced in October 1916 to a year in prison. Although the humiliating defeat at Caparetto in October 1917, where a young Captain Erwin Rommel won his *Pour le Mérite,* triggered his recall to service early in 1918 as the Director General of Aeronautics, the war ended too soon for Douhet to make a difference. He retired prematurely in disgust at his inability to influence Italian policy on the use of air power.

In the 1920s Douhet began publishing his ideas more widely. He was a regular

> **DOUHET's KEY POINTS**
> 1. There could be no effective defence against air power.
> 2. Air power allowed a nation to strike at vital centres of any enemy with impunity.
> 3. The real use of air power was as a bomber force that should not be squandered in support of ground troops.

contributor to *Rivista Aeronautica*, the official journal of the Italian Air Force. He published *Il Dominio dell'Aria* (*Command of the Air*) in 1921. With the Fascists' ascension to power in 1922, Douhet became a *cause célèbre*, regularly quoted in the Fascist daily *Il Popolo d'Italia*. Shortly after seizing power, Benito Mussolini appointed Douhet as Director General of Military Aviation. Nonetheless Douhet, while appreciating that the Fascists had created an autonomous air arm, continued to issue open critiques of Italy's aeronautic policies. Part of his criticism was based on what he believed to be the undue emphasis placed upon fighter aircraft, his personal conviction being that bombers were the chief component of air power.

Douhet was convinced that modern war was linked inextricably to mechanization, an idea that resonated with the Fascists. "Very much as Fuller had done, Douhet developed his historical interpretation of the growth of mechanical warfare to be understood in terms of the overall industrialization of Western society."[3] In *Dominio dell'Aria*, he declared unequivocally that victory in future wars would belong to whomever could achieve air superiority, that is, victory in the air. Douhet maintained that the advent of air power had extended battlefields to include civilian population centres including civil infrastructure. The result of this extension was that aerial bombardment could be used to weaken the civilian will to fight.

To Douhet, air power had to be used offensively: "Because of its independence of surface limitations and its superior speed – superior to any other known means of transportation – the airplane is the offensive weapon par excellence."[4] For him, it should be used to target civilian population centres: "[We] need only envision what would go on among civilian populations of congested cities once the enemy announced that he would bomb such centers relentlessly, making no distinction between military and non-military objectives."[5] Before he died in 1930, Douhet laid the foundation of air power's first coherent theory. His work can be summarised in three major points: Due to the offensive excellence of air power, there could be no effective defence against it; air power allowed a nation to fly directly over and then strike at the vital centres of any enemy with impunity; and the real use of air power was as a bomber force that should not be squandered in support of ground troops. Douhet's influence came mostly *post-mortem*. Not many people read his theories and a popular English translation did not exist until 1942. However, his theories were known in Langley, Virginia. The US Army Air Corps borrowed heavily from his work for their 1926 manual *Employment of Combined Air Force*.[6] To this day *Command of the Air* remains the theoretical cornerstone of all strategic air power theory and Douhet is still actively read and quoted.

Hugh Trenchard (1873-1956)

The father of the Royal Air Force (RAF) was not a theorist, but his influence built, expanded, and developed British air power. He became not only the Chief of the Royal Air Force, but also the custodian of British air power thinking. Having come to the concept of air power late in life after almost a full career in the infantry, initially he was firm in his convictions that air power's primary role was in support of ground forces and was opposed to the creation of an independent air force and as well as the idea of strategic bombing.[7] Throughout the First World War, as head of the army's Royal Flying Corps, Trenchard was completely committed to the employment of tactical air power. Deeply loyal to General Sir Douglas Haig, Trenchard's tactics of using aircraft for reconnaissance and ground support came at a staggering cost in terms of pilot casualties. Nonetheless he persisted.

Between the wars Trenchard was converted from supporting tactical air power to strategic air power. He came to value bombers and although he increasingly gave priority to bombing, in contrast with Douhet, he never completely gave up his belief that air forces had to work closely with ground and naval forces. Trenchard's influence was felt primarily through his capacity as head of the Royal Flying Corps and then RAF Chief of the Air Staff. Trenchard was not he was not a great communicator. As both taciturn and someone with poor writing and speaking skills, the air marshal was obliged to achieve his goals primarily by internal memos and orders.[8]

In the 1920s, with the war over and budgets tight, the RAF went looking for a role to save itself from the admirals and generals looking to 're-absorb' it. "Air Marshal Hugh Trenchard, RAF chief of staff, sought a mission that would justify the service independence of the RAF. The effectiveness of a few aircraft in putting down a minor rebellion in British Somaliland in 1919-20 provided Trenchard and the Air Staff the concept of an independent mission for the RAF."[9] The Colonial Office declared it a success: "The Royal Air Force had a peacetime mission: It could serve as Britain's frontier police force."[10] At the same time, Trenchard was adopting some of Douhet ideas and thus he became a conduit to through which the RAF came to believe that the aircraft would always get through, that the offence was the stronger form of war (contrary to Clausewitz), and that air superiority was a prerequisite for all warfare.

Trenchard's influence on the RAF was pervasive. As its head for its ten formative years, he established many of the policies which would later grow to become doctrine. Further, he was a close friend of Mitchell, having befriended him during the First World War. Trenchard went into retirement before the Second World War but lived to see the phenomenal growth of air power, dying 10 February 1956.

William (Billy) Mitchell (1879-1936)

Mitchell was born into a wealthy and influential family and was commissioned into the US Army Signal Corps. He was the first American to embrace Douhet's vision of offensive, strategic air power; he agreed with the Italian's vision of striking vital centres and did not see any moral or legal compunction against attacks on civilian targets, although he did stress the need to focus upon military objectives. Like Douhet, he argued that targeting civilian industry was more humane than the slaughter he had seen in the trenches during the First World War. Mitchell wanted air forces separated from the narrow constraints imposed upon them by both the army and navy. He saw air power's usefulness as strategic, not tactical. "Mitchell's persistent jibes at the Navy were especially nasty, and ... they not only fostered bitter interservice rivalry but also spurred the Navy to greater efforts in developing carrier-based aviation – the precise opposite of what Mitchell intended."[11]

In 1916 Major 'Billy' Mitchell, the erstwhile youngest member of the Army General Staff, was appointed to the head of the US army's aviation section. In early 1917, he was sent to Spain as a military attaché but was transferred to the Western Front with America's entry into the war in April. Initially, Mitchell was an air observer attached to the British and French air services. The British commandant, Trenchard, took an instant liking to the young American and the two became close and fast friends.

When General John 'Black Jack' Pershing arrived in France in June 1917 at the head of the American Expeditionary Force (AEF), Mitchell was made his chief of air staff. He immediately set about to convince Pershing of the requirement for a very large air force contingent. Pershing agreed and Mitchell was given the task of training and organizing it. Personally convinced of the promising role for strategic bombing in war, Mitchell was thwarted in proving his theories by the sudden end of the war. His personal drive and skill moved him quickly up the ranks, but immoderate language would soon be his undoing.

Within the United States, Mitchell became a popular figure but began to anger the service chiefs with his incessant promotion of strategic air power independent of the army and navy. In the early 1920s, he demonstrated that aircraft could not only hit ships at sea, but they could also sink them. Finally overstepping his bounds, like Douhet almost a decade before him, Mitchell was court-martialled in 1925:

> Not content to remain quiet, when the Navy dirigible 'Shenandoah' crashed in a storm and killed 14 of the crew, Mitchell issued his famous statement accusing senior leaders in the Army and Navy of incompetence and 'almost

treasonable administration of the national defense.' He was court-martialled, found guilty of insubordination, and suspended from active duty for five years without pay. Mitchell elected to resign instead as of 1 February 1926 and spent the next decade continuing to write and preach the gospel of airpower to all who would listen.[12]

Mitchell's views were prophetic but not new. He was a strong propagandist, completely committed to the idea of the establishment of a separate and equal US Air Force (USAF). His public outcries centred on the fact that splitting air power between the army and the navy made it both less efficient and less effective.

It would take another two decades before the attack at Pearl Harbor proved Mitchell right. Although some of his ideas diverged slightly from his Italian and British counterparts, most of this divergence was in practice rather than in conception. "There cannot be much doubt that Mitchell had an enormous impact on the foundation and development of the United States Air Force – and not just the man, but also the myths surrounding him."[13] Still held up today as the model for all aspiring USAF officers, he died in 1936 without seeing the devastation that air power could wreak.

John Boyd (1927-1997) and John Warden (1943-)

No discussion of modern air power can be complete without briefly mentioning two USAF colonels: John Boyd and John Warden. Both men have had major impacts upon not only their own military but also upon NATO. John Boyd's observations on aerial

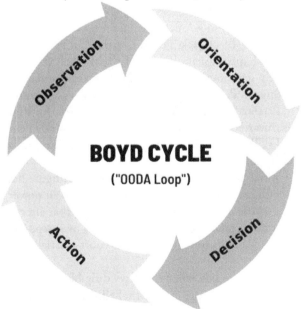

combat led him to write and study extensively on what he called decision-action cycles. Though he went on to design the F-15 and F-16 fighter aircraft for the USAF, and his more focused work on air power guided USAF thinking on achieving what has come to be referred to as strategic paralysis, his enduring influence is not confined to air power theory. Although Boyd does not offer a specific theory of air power his thoughts and writings on conflict have had significant consequences for the employment of air forces at all levels of war. Boyd's greater contribution was really in the creation of the underlying tenets, which led to Manoeuvre Warfare (MW), a modern and pervasive theory that is too broad to discuss here.

John Warden, who almost single-handedly wrote the air campaign plan for the 1990 invasion of Iraq, began with traditional air power theory based on Douhet's thoughts and then expanded his thinking to an almost evangelical zeal in his belief in the power of strategic bombing. Although Warden's theory is, strictly speaking, an air power theory, his ideas soon morphed into a new military theory – Effects Based Operations (EBO), which can be briefly defined as a theory of seeking to obtain a desired behavioural outcome as opposed to a simple destruction of enemy capability.

Based on air power, Warden's Five Ring Theory uses air power as but one part of a greater theory. Thus, Warden's ideas, like those of Boyd, are not really limited to air power theory. In Boyd's case, his influence has been pervasive, even if it has been outside the realm of air power. Boyd's studies on winning and losing in combat were precursors to the development of MW and his interpretation of military decision-action cycles is now embedded, to one degree or another, in the doctrine of Canada, Britain, the US and most other NATO nations. Warden's influence is somewhat more difficult to accurately gauge. Certainly, in the USAF, his ideas on strategic bombing

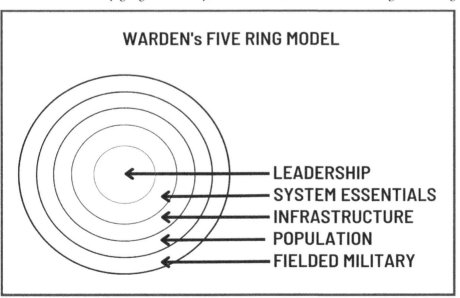

have reinvigorated the thinking of many serving officers. His contention that air power, when properly applied, can obviate the need for ground troops harkens back to Douhet and has, once again, ignited the debate in professional military circles over the proper use of air power in a joint environment. Nonetheless, it is in the arena of EBO that Warden's ideas are receiving their greatest attention.

The World Wars

Small wars often act as dress rehearsals for larger conflicts, and the Italo-Libyan War of 1911 was no exception. The Italians used aircraft frequently and to good advantage: "The Libyan campaign had taught the Italians, at least, the usefulness and the rapidity, as well as the reliability, of air reconnaissance of the other side of the hill; the need for accuracy in bombing; the dangers of ground fire; and the limitations of equipment."[14] From a near standing start during the First World War, it was not long before opposing pilots began exchanging gunfire in the air, which led to planes being armed with machine guns. The next logical step was the development of fighter aircraft that, naturally, required the development of fighter tactics. Planes fighting each other soon became a tactical activity all its own. Tactical air power theory was born.

The progress of air power during the First World War was particularly rapid. Without any theory upon which to draw, development was haphazard, *ad hoc*. This was both a new tool as well as a new form of warfare. Conservative generals and admirals, uncertain of what these new machines might bring to their respective services, were not ready to fully support their air arms at the cost of potential harm to their traditional power bases.

In a few short years, aerial bombing progressed from pilots dropping grenades by hand to the large-scale raids by zeppelins and Gotha bombers on London, ushering in the concept of strategic air power. Contrary to some predictions, the raids had little positive strategic effect. In fact, the opposite was true:

> Although of relatively limited military effectiveness, the most significant air power development of the war was to be that of strategic bombing. Targets were both military and civilian, and the greatest shock was to Britain, which, as an island nation, had long felt secure from enemy attack. As a consequence, the Zeppelin and Gotha raids on London had a disproportionate effect on future air power thinking. Some 9000 bombs were dropped by German airships and aircraft on Britain during the whole of the war, killing some 1413 people and wounding a further 3408. The British public clamour was for air defence at home and retaliatory strikes on Germany.[15]

Mitchell and Douhet observed the destruction and saw the untapped terror potential that mass bombing offered. This belief that aerial bombardment could force an enemy to surrender was a misapplication of the strengths and weaknesses of air power; both British and German civilians, for instance, proved admirably stubborn in their refusal to be deterred from their efforts because of strategic bombing.

The First World War had been fertile soil in which air power could grow, but the rapid growth had come at a staggering price. Without the time or inclination to methodically develop air doctrine, sound tactics, or strategies for the employment of their newfound capabilities, trial and error had been the default method of development for aircrews. Consequently, the casualty rates for the young pilots and their new air machines were staggeringly high. During the four years of conflict of the approximately 175,000 military aircraft built by the combined factories of Britain, Germany, and France, almost 117,000 of them were destroyed.[16]

The First World War, the 'war to end all wars,' had incurred a bloodletting of horrific proportions and the cessation of hostilities understandably saw an almost universal desire to decrease spending on all types of military expenditures. Armies demobilized and flotillas were anchored. If airplanes had been the technical marvel of the war, then they were also the expensive newest addition to the military arsenal, both in terms of blood and treasure. The RAF was quickly whittled down to a manageable and affordable couple of dozen squadrons. The same fate faced the Commonwealth, French, American and Italian fliers. As the declared losers, the Germans fell victim to the Treaty of Versailles, which outlawed military aircraft altogether for them.

During the conflict, there had been little penchant to consider what the future of air power might entail or, perhaps more correctly, no one had made the effort to publish the thoughts being discussed among the fraternity of new fliers. With the war over, time was available for reflection, theorizing, writing and most importantly, publishing the fruits of these intellectual labours. Douhet, and Mitchell extrapolated the limited experience of the First World War to claim that the bomber was the primary war winning capability of the future. In the immediate postwar period, they set out to publish their ideas promoting the idea that strategic air power and the bomber would change the nature of war for all time. The first and undoubtedly the more influential, was Douhet. His writings captured the imagination of strategic thinkers everywhere.

After 1918, air power languished – except where it was forbidden. The Germans struck a secret deal with the Soviet Union to train and develop this outlawed capability. The Nazis created the *Luftwaffe* in 1935, but unlike all other air forces, it did not espouse the philosophy of Douhet – except for the aspect of air dominance. German air power remained tactical and was focused on supporting land and naval forces. It was flying artillery. The inter-war years were problematic for air power theory, almost all proponents of strategic air power overstated their cases, whether it was the psychological effects of bombing, the ability to get through anti-aircraft defences or the devastation that bombers would cause to their civilian, military, or industrial targets. Technical proficiency was declared; accuracy that would not be realized for many decades was claimed.

STRATEGIA

From the perspective of air power, the two most important events between the wars were the Manchurian Crisis (1931-32) and Spanish Civil War (1936-1939). In the Far East, the Japanese Air Force used dive bombers to support lightning ground force strikes against the Chinese and since there was little resistance, the Japanese quickly gained air supremacy and the planes operated freely. The more important event was the Spanish Civil War. This conflict saw action by air forces from Spain, the USSR, Germany, and Italy. In fact, the *Luftwaffe* used the war as an experimental laboratory where they developed, tested, and exercised to great effect their emerging doctrine of joint operations in support of ground forces. Concurrently, they were also able to develop new tactics: battle formation flying quickly replaced tight display formations; air power was integrated with ground operations using forward air controllers, effectively turning dive bombers into long range artillery; strategic bombing for maximum effect was tested as were special munitions, the most infamous being the firestorm created in the town of Guernica.

The devastation of this Spanish town was a foretaste of the future, and its bombing arose from debate over which targets should have precedence. Douhet's theory demanded attacking civilian targets but both British and American air power proponents argued that the focus must be on military ones. In the end, both types of targets were hit. Guernica was the trial run and years later the pictures of London, Coventry, Hamburg, and Dresden after their bombings would be held up as proof of the ability of air power to wreak total devastation. Nonetheless, neither the loss of materiel, nor the terror inflicted on civilian populations, nor the strategic surprise of attacks like Pearl Harbor became the bases of war-winning strategies. The effects achieved were neither lasting nor were they conclusive.

Unlike the First World War, the Second saw air power ready to play an important role. During the six years of war, practically every air power theory was tested, adapted, and employed. Necessity forced the advent of new methods, technologies, and techniques. Advances in aviation made practically every civilian centre susceptible to attack. But as so often before, the promises of technology were not fully realized and just with all other arms races, each new technological development was quickly matched by a counter advancement. Of all the belligerent European powers, only the *Luftwaffe* did not come under the intellectual sway of Douhet. Built and commanded by Göring, himself a First War air ace, it had always been and remained focused on the objective of destroying enemy military capability: aircraft, ports, and airfields. So, although the *Luftwaffe* practised some strategic use of air power, its focus was overwhelmingly tactical. The strategy was working and by early September of 1940 Göring was convinced that England was ready for the *coup de grace*. So, the *Luftwaffe* switched its emphasis from tactical use to strategic, from military to civilian targets, convinced that terrorizing London would set the stage to invade the island. But it was too little too late; the breathing space created from changing targets allowed the RAF to recover and rebuild. The invasion never came.

WAR IN THE AIR

The Second World War offered an important lesson in air power: control of the sky was key to victory, but it was not a guarantee. The Allied air supremacy over France during the Normandy campaign allowed extensive reconnaissance, defence against enemy air attack and almost a complete disruption of German logistics. And yet, the Germans established a punishing defence that lasted for over a year. Air power needed to be closely coordinated with ground forces – no simple task. Most importantly, air forces could not hold ground.

Despite the Douhet-style terror achieved against Hiroshima and Nagasaki, the end of the Second World War made the disappointing verdict clear:

(1) Precision bombing did not work; bombing necessarily was a brute and indiscriminate instrument, and the degree of bombing necessary to make a difference required the barbaric slaughter of civilians on a mass scale. (2) Not even bombing on the barbaric level could in itself suffice to horrify an enemy into surrender. (3) Even if such bombing could compel surrender, it still failed to deliver on the revolutionary promise of victory without the price of combat; bombing was just another form of combat – indeed, one with unusually high mortality rates. (4) Atomic bombing could deliver on air power's promise of victory through terror without combat, but it was so awful that no one ever wanted to use it again. War remained fundamentally a contest to be decided between fighters on the ground, and the generals' dream remained a dream.[17]

Clear now perhaps, but it was not so clear at the end of the war. The threat of nuclear bombardment defined an entire era and kept the world conflicts from engulfing the globe in another world war. From the nuanced difference between US and British doctrinal definitions, we can already begin to see the influence of nuclear weapons.

DEFINITIONS OF AIR POWER

BRITISH: The ability to project military force in air or space or from a platform or missile operating above the surface of the earth.

AMERICAN: The synergistic application of air, space, and information systems to project global strategic military power.

STRATEGIA
Cold War and the Nuclear Age

The Korean War (1950-1953) proved that conventional war remained possible after the advent of nuclear weapons. Jet aircraft fought each other in large numbers for the first time; the need for joint operations was re-learned. Despite the introduction of helicopters and the primacy of a ground war, the USA and USSR expanded their air forces and all the world's major powers continued to base their air strategies on bomber fleets – in particular, fleets that could carry nuclear weapons. In America, Generals Hap Arnold and Curtis Le May shaped and built massive fleets of bombers. Simultaneously, the Americans, followed by the world's major powers, began building tactical as well as strategic missile forces. This policy was the strategic offence, which was to act as the strong deterrence and therefore a strategic defence. The Soviets, British, French, and Chinese soon followed the enormous build-up of strategic nuclear forces. Thus, it was there that resources were channelled.

In many ways, the Cold War was a unique period in modern history. The rapid development of air power that became so closely linked to the development of nuclear weapons had an unexpected consequence: Western military thought and theory took a hiatus of sorts. Humanity became mesmerized by the destructive potential of nuclear weapons. Consequently, classical military theories and their political theory counterparts were overthrown in large part by the belief that nuclear weapons had relegated most of what these theories offered to the dustbin of history. The new and unimaginably destructive weapons convinced many that classic theories had nothing to offer. Western society came to accept the notion that by virtue of their destructive power, nuclear weapons had somehow changed the nature of war, when once again, what had changed was only its character.

The premise appeared legitimate. The devastation wreaked upon the Japanese mainland in 1945 was something that had never been seen before. It was, therefore, appropriate to consider whether new theories were needed to create new strategies. Using the model presented in the 'The Unknown Nature of War', someone would have had to go back to first principles and review any underlying theory (and possibly even philosophy) to confirm its validity and relevance. Certainly, some did so. The study of classic military and political theories was not abandoned. This study was simply given less credence than the much-touted belief that war had forever changed.

What took the place of the classical processes was military strategy that was generated from philosophies having little to do with war. During the Cold War, nuclear strategies were repeatedly formulated without benefit of relevant military theories. Instead, they stemmed from mathematical, sociological, or economical models and theories. Initially, nuclear strategy grew out of classic air power theory, primarily the thoughts of Douhet, a theory that had only proven itself to be marginally successful. Strategic bombing did demonstrate its utility but not nearly as effectively as the air power theorists had predicted. In the closing days of the Second World War, however,

WAR IN THE AIR

Douhet's apostles got a new lease on life. The limitations that they claimed had held them back were wiped away by the power of atomic weapons. The utter destruction of Hiroshima and Nagasaki were proof – or at least that was the claim. The use of atomic and then nuclear weapons soon came to rule much of the strategic thinking for the following half-century. Air power appeared to offer the ultimate guarantee of victory:

> Initially, when atom bombs first made their dramatic entrance onto the international stage, they were discussed and understood in terms derived from the established theories of airpower ... New concepts and approaches developed in an attempt to come to terms with a situation in which a war in which the most formidable weapons available were used would, in all probability, be catastrophic for all concerned.[18]

While political leaders pondered what this meant and military leaders raced to build ever-larger fleets of nuclear capable aircraft, academics grappled with what effect such destruction had on the formulation of national strategy.

In the US, the RAND Corporation (Research and Development Corporation, a Santa Monica, California non-profit institute funded by and with close ties to the USAF) had a great influence in the development of air power thinking. Deterrence theory, predominant at the time, required large fleets of bombers that were widely dispersed. At the same time missiles were coming of age. The combination of missiles and aircraft was deemed to be more effective than aircraft alone. As the Cold War deepened, the threat of flexible strategic nuclear retaliation kept an uneasy peace – demonstrated in the Cuban missile crisis of 1962. Eventually, however, American involvement in Vietnam and the Soviet intervention in Afghanistan came as harsh reminders of the limitations both of tactical and strategic air power. In both wars, air power theorists were once more taught the lesson that conventional bombing did not work the way that the theories promised.

The Vietnam War was a textbook case of the failure of strategic bombing to influence the outcome of a ground war:

> [A]t the core of the strategy there was a *theory*, and this *theory* was a revival and a refinement of the World War II dream of victory through bombing. The *theory* was called by various names, "graduated response" and "phased escalation" among them, and it held that a calibrated and predictably increasing use of bombing against North Vietnam would eventually force the Communists to abandon their efforts to take over South Vietnam and to accept a negotiated peace ...[19]

If we take 1964 as the starting point, the figures are startling. By 1968 the USAF was flying about 12,000 missions a month to little avail. That same year, President Lyndon

Johnson ordered a halt to the bombing campaign, Operation Rolling Thunder. In its four years, the operation had seen some 300,000 missions and almost three quarters of a *billion tons* of bombs dropped. The North Vietnamese were not cowed. Eventually the Americans accepted defeat, even if the air power advocates did not, insisting that the use of air power had had political constraints placed upon it and that if it had been unfettered it could have been decisive.[20]

Vietnam was not an isolated case. Throughout the Cold War smaller conflicts always brought forth air power enthusiasts ready to argue a reinterpretation of Douhet's theories, providing justification for a broad array of capabilities. Even so, the potential NATO versus Warsaw Pact confrontation in Europe continued to be the scenario that governed force structures and technologies. The binary policy of massive retaliation grew into NATO's espousal of the more nuanced flexible response that combined possible nuclear retaliation with a series of conventional options as well as the use of tactical nuclear weapons. Each new conflict brought new lessons as well as a renewed belief in bombing. The Korean War was the first sizeable demonstration of the value of helicopters on the battlefield; the Yom Kippur War demonstrated the need for tactical fighters and the value of air defence; the Vietnam War demonstrated the effectiveness – or ineffectiveness – of strategic bombing; the Falklands War demonstrated the need to maintain carrier-borne fleet air arms; and the Russo-Afghan and the Russo-Chechin Wars demonstrated the vulnerability of battlefield aviation. As always, proponents saw vindication while opponents saw failure. Whatever the arguments, each of these wars was seen as an isolated incident and the major world powers continued to focus most of their interest and money on the massive air fleets required of the NATO-Warsaw Pact standoff.

The military airpower enthusiasts were not alone. In the words of Professor Lawrence Freedman, there was:

> ...a vast outpouring of books, articles, papers and memos from civilian representing many academic disciplines and often organized into research institutes concerned with few things other than the problems of national security. Their writings were replete with new and arcane concepts, which sometimes served to clarify but often only obfuscated, and were caught up with a forbidding miasma of acronyms and jargon.[21]

However, most of these thoughts were not about military theory or even air power theory; they were about the technology of how to use the weaponry. In other words, the debate was not one of air power theory as much as they were about warfighting. During the Cold War, the focus of air power thinking became ever more concerned with the delivery of strategic nuclear weapons. The drive for technological advancement and superiority blinded the air power community to the need for military strategy that was grounded in one or more military theories. Having been beaten into space in 1957 by the USSR the US became increasingly determined to build and maintain a credible deterrent capability. This determination was echoed in the USSR. The technological arms race soon outstripped the logic that triggered it in the first instance and the resources required to maintain a technological superiority created its own *raison d'être*.

There is an historical analogy that is instructive: In 1941, Hitler sent a relatively junior but tactically brilliant Erwin Rommel to Africa with a simple operational directive: 'Tie up as many British forces as possible to give us a free hand in Europe.' What Rommel did was just the opposite. His battlefield victories added legitimacy to his constant demands for more troops, eventually causing the German General Staff to become seriously concerned about how much Axis strength was being diverted to a secondary theatre of operations. Drawing so many troops to Africa eventually allowed for the invasion of Italy.

In a similar way the post-war focus for the victorious powers was on the development of nuclear weapon delivery capabilities. For the US, USSR, – and to a limited extent France and the UK – the key role of air power was nuclear weapons delivery. Le May built a vast strategic bomber fleet that far exceeded the capability of any other nation to ensure that nuclear deterrence theory constructs could work in practice. It became a cyclical and self-fulfilling argument: greater resources were needed to ensure that nuclear weapons would work; the greater the resources allocated, the more important the nuclear weapons became. In other words, the theory (or what passed as theory) became less important than the strategies and procedures, which stemmed from it. More bluntly, theory was devalued.

The unforeseen and rapid escalation of the Cold War caused a certain urgency in the requirements for strategies in the two alliances facing each other in Europe. The military strategies that developed necessarily grew out of classical military theories and political strategies. But as the Cold War progressed from one stage to the next the strategies soon became disconnected from their classical roots. They soon became more fixated on technological changes being made in nuclear weapons rather than any underlying philosophical, political, or theoretical fundamentals. It was widely believed by RAND Corporation analysts such as Thomas C. Schelling, Herman Kahn and the extremely influential mathematician Albert Wohlstetter that classical theory was not suitable to the task of determining strategy for nuclear warfare. But it was really their followers who broke the links between classical theory and nuclear strategy. By the mid 1960's, Cold War strategy had become fully disengaged from classical theories. "Specialists whose starting point was the extreme qualities of nuclear weapons rather than the timeless qualities of international politics" controlled its creation.[22]

Once released from its historical tether, nuclear strategy was hijacked by technically oriented specialists. It was only a matter of time before the fixation upon the destructiveness of nuclear weapons led to strategies that were dependent upon experts who concerned themselves with weapon accuracy, range, yield potentials, kill probabilities, targeting and a myriad of other technical domains. Graphs and matrices comparing equivalent mega tonnages of bombs or missiles, the probability and levels of destruction and survivability of initial and secondary 'nuclear exchanges.' Mathematics usurped diplomacy; strategy became the stepchild of science instead of the product of classical military theory. Gradually, almost imperceptibly, strategic requirements assumed primacy over political imperatives.[23] The aptly named strategy of mutual assured destruction (MAD) was the result. In an ironic twist, the belief by the nuclear specialists that the 'balance of terror' had to be maintained strengthened the need to understand classical military theories: the classical theories depended upon understanding that political instability could trigger a war independently of whatever inherent stability nuclear deterrence created internationally. Although not understood at the time, the abandonment of classical theories made them invaluable. In the meantime, mathematicians and economists drove nuclear strategy.

Two such men were the deterrence theorists Thomas C. Schelling and Robert J. Aumann. Independently, but in tandem, they explored arcane game theory to derive required force structures to maintain deterrence as well as keep the upper hand against potential opponents. These two research scientists created the foundations for all future work in non-cooperative game theory as well as its application to problems and major questions in the social sciences. Consequently, they shared the 2005 Nobel Prize in economics 'for having enhanced our understanding of conflict and cooperation through game-theory analysis'. Each approached the subject from his own perspective: Aumann from mathematics; Schelling from economics. But their aims were the same. Both saw the potential for game theory to be a useful theoretical tool that could be employed in the analysis of human interaction.

Although trained as an economist, in his 1960 book *The Strategy of Conflict* Thomas Schelling set forth his vision of game theory as a unifying paradigm for the social sciences. He focused on games that were not 'zero-sum.' Instead, he put emphasis on the fact that almost all multi-person decision problems contained a combination of both differing and shared interests, some individual, some communal. He felt that the interplay between the two types of interests could be analyzed by means of non-cooperative game theory. Schelling began his work with the concept of 'bargaining.' Prior to him, economists invariably began with the premise that the object of all interactions was a fair outcome. Schelling took a more realistic approach using, for instance, the example where two trucks met on a road where only one could fit. He then expanded the example to a military scenario.

Imagine two countries undergoing tensions over a disputed territory. Each could choose to mobilize its military. If both mobilized, then there would be a high probability of war, while the probability of a peaceful agreement about division of the territory would be low. Let the expected payoff to each country be zero if both mobilize. If instead, both countries refrain from mobilization, then a peaceful agreement over the territory would have a high probability, while the probability of war would be small. In this latter case, each country would obtain a positive expected

WAR IN THE AIR

payoff 'b'. However, if only one country mobilizes, it could take complete control of the territory without war, and neither the other country nor any other party could force a military retreat by the occupant. The aggressor would obtain payoff 'a' while the loser's payoff would be 'c', where a>b>c>0, war thus being the worst outcome. This overly simple "mobilization game" could be described by a payoff probability matrix, where one player (here a country) chooses a row and the other simultaneously chooses a column, with the row player's payoff listed first in each entry but we will not go down that path. Suffice to say that such mathematical interpretations seem to intrinsically deny human interactions in war, a domain filled with emotion. That said, such simple premises led to technical aspects that were manageable. They made complex situations more understandable, or so it seemed.

Robert Aumann's part in the development of game theory promoted a unified view of the multidisciplinary domain of strategic interactions. His work combined many seemingly disparate disciplines: economics, political science, biology, philosophy, computer science and statistics. Rather than using discreet models to deal with issues such as deterrence, perfect competition, or oligopoly, Aumann developed universal methodologies, which helped clarify the internal logic of game-theoretic reasoning – at least to some. Part of his work helped shed light on the difference between short short-term and long-term interactions and how they applied to strategy.

The simplest example of such an interaction is illustrated in the famous prisoners' dilemma. Let us imagine that Bob and Gord have been arrested for a robbery. They are isolation in separate cells. Each prisoner cares more about himself than about his accomplice. The Crown Attorney makes the same offer to each:

> "You can confess to me or remain silent. If you confess and your partner stays silent, I will drop all charges against you. Then, your testimony will be used to convict your accomplice. But if your accomplice confesses and you remain silent, he will go free, and I will convict you. If you both confess, I will convict both of you, but I promise to intervene with the Magistrate to ensure you both get a shortened sentence. If you both refuse to talk, I will be forced to settle for convictions on lesser charges of firearms possession. If you decide to confess, you will have to inform the guard before I come back tomorrow morning."

In this two-person 'game', each player has two potential strategies: to cooperate [a] or confess [b]. But the players are forced to choose their strategies simultaneously. Clearly, each prisoner's leading strategy is [b] for the simple reason that it is an optimal strategy irrespective of what the other prisoner does. But both players gain if they both play [a]. When played once (as in a nuclear war), the game admits only one solution that is known in probability theory as a 'Nash Equilibrium'. It is the situation where both prisoners choose strategy [b]. But the equilibrium outcome is worse for both players than the strategy pair where both decide to choose [a]. However interesting, we are again in the realm of mathematics and probability as with the situation above (a>b>c>0) and straying from classic forms of war-centric military theories and unlike in games, nuclear powers would only get to 'play' once.

Like Schelling's original work, this too made some of the esoteric mathematics more understandable. But it did not take long for the simple logic equations and matrices to evolve into obstruse and impenetrable equations that were wholly indecipherable to any but game theorists or advanced mathematicians. No doubt these equations were fascinating to many people. They were not, however, even remotely related to any war-centric military theory. For example, the two authors offer a scenario paraphrased below:

> Suppose Country 1 can commit to an infinite range of probability $\pi \in [0,1]$ of retaliation if Country 2 mobilizes. If Country 2's preferences are predictable then deterrence requires that $b \geq (1 - \pi) a$ or, equivalently, that $\pi \geq 1-b/a = \pi^*$. If we allow θ be the probability that Country 1 attaches to the possibility that Country 2 prefers to mobilize regardless of the retaliation threat. For $\pi \leq \pi^*$, Country 2 will still mobilize for sure, so the payoff to Country 1 is then $(1 - \pi) c$, a decreasing function of π. For $\pi \geq \pi^*$ its expected payoff is $\theta(1-\pi) c + (1- \theta)b$, again a decreasing function of π. Thus, deterrence (choosing $\pi = \pi^*$) is optimal for Country 1 if and only if $\theta(1-\pi^*)c + (1- \theta)b$ is at least as large as the payoff c from not retaliating ($\pi = 0$), or, equivalently, if and only if $\theta \leq (1 – c/b) /(1 – c/a)$.

I do not pretend for a moment to explain the scenario above, for what Shelling and Aumann offer is not strategy, *per se*. It is obviously advanced mathematical probability theory. Clearly, supporters of such complex mathematical military theories had never heard of French Admiral Hyacinthe Aube or his *Jeune École* and the havoc they had wreaked upon France.

The RAND Corporation was greatly influential in the development of Cold War air power thinking. The logic that deterrence depended on the nation being able to ride out a first nuclear attack, and yet still launch a devastating retaliatory nuclear attack became widely accepted. This led to a requirement for large nuclear forces widely dispersed and a comprehensive national air defence system. Yet even as the bomber forces were being built, another technology was offering more assurance of a second-strike capability. Offensive missiles, both 'air-breathing' and rocket-powered. In 1959, Bernard Brodie argued in a RAND study that inter-continental ballistic missiles could provide greater assurance of nuclear retaliation than manned aircraft; but that a mixture of systems was even better. In the US, Robert McNamara, an ex-US Army Air Force officer and Harvard Business graduate, brought to the post of Secretary of Defense in 1961 a keen analytical mind. He started the trend of thinking in capability terms when making investment decisions, a business formulation that continues in departments of defence to this day.

The fall of the Iron Curtain in November of 1989 and the unexpected collapse of the Soviet Union less than two years later changed the array of strategic players. The subsequent instability brought on by that collapse quickly made the demand for these nuclear specialists mostly superfluous. Inversely, the need to understand classical military theories and their resultant strategies based on classical political philosophies and theories once more came to the fore.

Political or moral aims should form the basis of, and guide the formulation of, military strategy. Edward A. Kolodziej, a research professor in the Department of Political Science at the University of Illinois, points out that this was not the case during the Cold War: "Even a casual examination of the evolution of nuclear strategic thinking and practice by the United States and the Soviet Union during the Cold War leads one to the conclusion that this fundamental principle was honored more in the breach than in the observance by both sides."[24] Cold War strategy demonstrated a simple human failing. What many strategists tended to do was to become enamoured of their own analyses of what force was needed to compel an opponent to behave a certain way. Unfortunately, what was too often overlooked was that opponents were human beings and not equations in some mathematical model. Specifically for this reason it was paramount that, to paraphrase Clausewitz, strategy be the servant of politics and not vice versa:

> The ascendancy of strategy over politics during the Cold War was also aided by the dubious claim of some planners that their theory of deterrence had reached the level of a scientific truth. Strategy was something to be left to experts. Only those possessed of the requisite and sanctioned language, conceptual tools, specialized language, and advanced computational skills merited entry into this discipline and dismal science. Only they might speak authoritatively about strategy and the placing of 'ordnance on target.'[25]

Military strategy was thus to be left to the experts. The difficulty was that the experts had little or no understanding of classical military theories.

By the end of the Cold War, air power systems had developed beyond what even Douhet's fertile imagination had foreseen. Satellites permanently orbited the globe, providing imagery of almost anything on the planet's surface; they also enabled worldwide communications. Airborne technology had made it possible to strike virtually any target on the earth's surface. Undreamed of accuracy allowed commanders to strike targets that were continents away in almost any weather and with practically pinpoint accuracy. Aircraft could shoot down other aircraft that they would never even see. Nonetheless, air power theory was still either ignored by those who were mesmerized by the technology, or it was fundamentally unchanged from what Douhet, and Mitchell had said decades before. The Cold War was over, but as the world entered what some called the 'air power decade' the theoretical edifice of air power was now much more questionable.

> This separation of nuclear strategy and politics during the Cold War resulted in bad theory and unfounded expectations about how states actually behave, wasted enormous human and material resources, ran unnecessary risks and needlessly courted disaster, and encouraged morally reprehensible nuclear postures by both sides. ... We tend to confuse and confound scientifically demonstrated knowledge about the destructiveness of nuclear weapons – which we know a lot about – with the paucity and internally flawed understanding of how humans behave when possessed of nuclear weapons. Simply put: Cold War strategists and social scientists, as would-be emperors, were without clothes – or at best they were scantily attired.[26]

In the end, most of the strategies and theories developed by RAND alumni and their government masters, were demonstrated to be intellectually bankrupt. Before his death, Bernard Brodie admitted that he believed that most of what was produced by his old firm was based on careerism and was unworkable. Both Robert McNamara and Henry Kissinger joined Brodie in reverting to similar beliefs.[27]

The Air Power Decade

When the Warsaw Pact disintegrated and the USSR dissolved, NATO nations raced to spend their 'peace dividends.' However, in August 1990, amidst the world celebrations and talk by academics of an 'end of history', Saddam Hussein invaded Kuwait. Six months later the US led a coalition of 29 nations in a war of liberation beginning with a massive air campaign that lasted almost two months. Once the enemy had been 'softened,' the ground invasion, expected to last weeks if not months, commenced. It lasted one hundred *hours*. Amongst the general jubilation one group of voices rang out: Douhet had been vindicated they said.[28] Air power had not only won the war; they had demonstrated that all future wars could be won this way, with ground forces merely going in after the fact to clean up and maintain the peace.

It was hard to convince the US government that the air power proponents were wrong. Decades of civil-military relationships wavered. The Navy, Army and Marine Corps were all wrong according to the Air Force.

> Decades later, some American airmen and airpower theorists would look to the British colonial example of air control as proof that airpower alone, or with just a handful of police and special ground units, could maintain peace and order over a large region. If the idea of air control was simple, cheap and

attractive in the 1920s, it was even more popular in an era when American forces where required to intervene and keep the peace under chaotic and complicated conditions in places such as Haiti, Somalia, Bosnia, Kosovo and elsewhere.

The only problem with the air control concept is that it was a myth.[29]

Having just demonstrated a high-tech 'sanitary' war-winning strategy, American policy shifted from concern over mass casualties to the loss of a single solder. The loss of a handful of soldiers in Mogadishu in October 1993 caused the US to abandon the mission. Casualty tolerance dropped to zero. Then came the crisis in the Balkans as the Republic of Yugoslavia literally came apart.

Orthodox Serbia was repressing the ethnic Muslim Albanian community. The UN and the Organization for Security and Cooperation in Europe (OSCE) tried and failed to bring peace. For the first time, NATO agreed to use military force in the absence of a UN resolution. The instrument of choice was strategic air power. It held the promise of success without casualties. Beginning on 24 March 1999, air power pounded the Serbian leadership to 'persuade' it to stop its repression. Almost 40,000 missions were flown with tens of thousands of bombs dropped. For almost three months, Serb positions were relentlessly bombed without a single NATO casualty. In June, a peace settlement was signed. But was it a pyrrhic victory?

> A lot of people in the Pentagon will tell you, for example, that the American-led ouster of the Serbian Army from Kosovo proves what glorious things you can do with air power and precision weaponry. [General] Bill Owens, a former vice chairman of the Joint Chiefs of Staff ... argues that Kosovo was a litany of failures. The precision-bombing campaign destroyed the Serbian industrial base, but it was not that great at hitting the Serbian Army, which managed to conduct a sadistic ethnic purge of Kosovar Albanians while under American fire. One weapon that might have worked against the Serbs, a fleet of Apache helicopters sent to Albania for just that purpose, stayed grounded in part because the Army and Air Force couldn't agree which service would be in charge, in part because the Clinton administration could not bear to put such a valuable weapon in harm's way.[30]

The most powerful nations in the world bombed an economically insignificant province, in a small Balkan state, destroying most of its infrastructure. Still, the Serbian government of Slobodan Milosevic did not alter its policies; his army lay intact. With no NATO troops on the ground, Serb combat units remained in their safe hideouts. Arguably this evidence suggested that it was not air power that brought Milosevic to the table. A small militia force, the Kosovo Liberation Army (KLA) forced Serb forces to move and become vulnerable to NATO aircraft and only when his military forces began to be destroyed by aircraft did Milosevic come to the bargaining table. Without the KLA, planes could have pounded Serbia indefinitely.

Whether the bombing 'won' the Kosovo campaign remains moot. From an historical perspective there has not yet been time to digest the results. Nonetheless, the point is that air power advocates continue to make the same types of boasts made by Douhet and Mitchell a century ago. Even official doctrine makes such claims. Current USAF doctrine presents aerial warfare as the ultimate example of Manoeuvre Warfare and lays it out clearly in *Air Force Doctrine Document 2 (AFDD 2), Organization and Employment of Aerospace Power*, USAF's capstone document. MW has lately been replaced by Effects Based Operations, a USAF interpretation of this theory at the strategic level. Using the theories of John Warden, the USAF renewed its desire to convince both political masters and the remainder of the military community that air power could lead the way in a new paradigm of war.

> Its *means*, as in Douhet's original vision, comprise the use of the air as a manoeuvre space to avoid enemy strengths. Furthermore, its means are fast, emphasize intelligence, deception and flexibility, and they can be used to strike directly at the enemy centre of gravity. Its *ends* are the attainment of victory by demoralizing the enemy and destroying his will to resist. Perhaps most importantly, as described in theories of strategic paralysis by John Warden, many air power advocates, while paying lip service to the synergy of all arms in the joint campaign, believe that air power should predominate in plans to defeat the enemy.[31]

The USAF heralded Warden's theories as the latest hope to regain both credibility and ascendancy. But the US Marine Corps and the US Navy refused to buy in; the US army was skeptical, and the US Joint Staff was inclined to agree, eventually dropping EBO altogether.

The search by air power advocates to convince others of the ability of air power to win wars continues unabated. But "technology-inspired panaceas which provide ad hoc tactical success may do so at the expense of long-term strategic flexibility."[32] For nations, the achievement of political will can only be won if the strategies it produces is a product of the *total* combined abilities of a nation's power and not just the employment of force. Recall Echevarria's comment regarding Germany's confusion of *power* and *force*.

Summary and Conclusions

Air power theory may be a latter addition to the greater Military Theory domain, but it has made extraordinary inroads. From its outset, the creation of air power theory has

WAR IN THE AIR

sought to displace land power and sea power theory as 'first among equals.' Douhet, Mitchell, and even Warden have claimed that air power could fundamentally alter the nature of war, that it could play not just a supporting role, that it could be *the* predominant tool of war. This claim is really a revision of the concept of naval blockade, of strangling an enemy through starvation, fear, and industrial incapacitation, come of age. Further, during the Cold War, air power theory became the basis upon which much of nuclear strategy was based and thereby predominant.

The development of flying machines took warfare into the third dimension in a manner previously thought impossible. The advent of air power brought theorists who mistakenly claimed that aircraft had altered the nature of war: acting alone, without land or sea power, nations could now win wars using only air power. Ostensibly the air battles of the First World War were fought without any underlying theory. Douhet filled the gap in 1921 and the interwar years would see him, and Mitchell spread the gospel of strategic bombing. In Britain Trenchard, originally a strong proponent of tactical air power turned slowly away from his unwavering support of the army to become a bombing enthusiast. In Germany, Douhet was not fully espoused; Göring focused his *Luftwaffe* on tactical support to the armoured warfare that proved successful on the battlefield. *Blitzkrieg* demanded that air power be tied to ground forces and, despite the near-run Battle of Britain, the fettering of air power to land power remained a touchstone of German doctrine. In America, Mitchell's strident claims and publicity seeking forced the generals and admirals to court-martial him but, in the process, he demonstrated the need for both aircraft carriers for the navy and strategic bombers for the army.

The Second World War saw the advancement of both tactical and strategic air power and, while there can be little doubt that air power was important in winning the war, the prophesies of Douhet and Mitchell were not realized. Experience demonstrated that even though control of the air was becoming a necessity for success on the ground, ultimate victory could only be won on the ground. Meanwhile, prophets of air power continued to fixate on strategic bombing, which finally fulfilled its deadly promise with such devastation, in Hiroshima and Nagasaki, as to become practically self-defeating.

After 1945, all military development took place under threat of nuclear war between NATO and the Warsaw Pact. In the beginning, both sides developed more and more sophisticated weapons that could be delivered by aircraft. Aircraft then gave way to unmanned missiles. In many ways, the development of air power during the Cold War kept tensions high without allowing them to spill over into a shooting war by ensuring that any conflict would result in mutual destruction. Such a conflict would be unwinnable. Unfortunately, in the process of developing deterrence strategies, the strategists fell so deeply in love with the means that they had devised for waging war that they became blind the relevance of these means to actual war.

Precision bombing got a new lease on life with the Gulf War and Kosovo, deservedly or otherwise. With modern warfare a fully three-dimensional endeavour, any successful application of military force must include control of the air. But such power must then be coordinated with land and naval forces. Air power is a high-tech endeavour, which presupposes an industrial base as well as skilled manpower. Air power is not cheap, nor can it seize nor hold terrain. But under limited circumstances, particularly in short conflict scenarios, air power can provide political options not offered using land or naval power alone.

Although still in its infancy, war in the air made startling advancements since its inception at the dawn of the 20th century. The development of war in the air has easily been the most rapid of all types of warfare. The early apostles were mostly wrong, although they still have their proponents amongst those looking for quick, clean, mechanical, solutions to problems that have plagued mankind for centuries. Too often, though, they overstated their cases – at times coming close to declaring that air power was indeed the Holy Grail of war, the ultimate decisive weapon. Last, it should be appreciated that air power has, in subtle ways, altered the civil-military relationship. Where in the past politicians were obliged to rely upon their generals to keep them informed of how the political objectives were being achieved, the 20th century saw more than one occasion where political power was able to interfere directly with operational plans.

Notes

1. Robin Higham, *Air Power: A Concise History*, (New York, NY: St. Martin's Press, 1972), 2.
2. Giulio Douhet, *The Command of the Air*, Dino Ferrari, translator, U.S. Air Force ed., (Washington: DC, 1983), 140.
3. Azar Gat, *Fascist and Liberal Visions of War: Fuller Liddell Hart, Douhet, and other Modernists*, (Oxford: Clarendon Press, 1998), 66.
4. Ibid, 15.
5. Ibid, 20.
6. James L. Stokesbury, *A Short History of Air Power*, (New York, NY: William Morrow & Co., 1986), 128.
7. Mets, David R. Mets, *The Air Campaign: John Warden and the Classical Airpower Theorists*, (Maxwell AFB, AL: Air University Press, 1999), 21.
8. Ibid, 21-2.
9. James S. Corum, "The Myth of Air Control: Reassessing the History," Aerospace Power Journal, (Winter 2000), 61.
10. James L. Stokesbury, *A Short History of Air Power*, (New York, NY: William Morrow & Co., 1986), 124.
11. Phillip S. Meilinger, *Biographical Sketch of General William "Billy" Mitchell*, available at http://www.airpower.maxwell.af.mil/airchronicles/cc/mitch.html. Note: this is an archived page and must be accessed using an Internet Archive service.
12. Ibid.
13. David R. Mets, *The Air Campaign: John Warden and the Classical Airpower Theorists*,

(Maxwell AFB, AL: Air University Press, 1999), 42.
14. Robin Higham, 22.
15. Timothy Garden, "Air Power: Theory and Practice," available from http://www.tgarden.demon.co.uk/. Note: this is an archived page and must be accessed using an Internet Archive service.
16. Ibid.
17. Michael Kelly, "The Air-Power Revolution", The Atlantic Monthly, (April 2002), 20.
18. Lawrence Freedman, *The Evolution of Nuclear Strategy*, 3rd ed., (New York: Palgrave Macmillan, 2003), xix.
19. Michael Kelly "Slow Squeeze," The Atlantic Monthly, (May 2002), 20. (Emphasis added.)
20. Robin Higham, 279.
21. Freedman, xvii.
22. Freedman, 459.
23. Edward A. Kolodziej, "Lessons of Hiroshima: A Geopolitical Perspective," Swords and Ploughshares, (Spring Summer 1995), 22.
24. Ibid, 21.
25. Ibid.
26. Ibid.
27. Gwynne Dyer, *War: The New Edition*, (Toronto: Random House, 2004), 322.
28. Major Kurtis D. Lohide, "Desert Storm's Siren Song", Airpower Journal (Winter 1995).
29. James S. Corum, "The Role of Airpower in Future and Current Small Wars," Centre for Defence and Security Studies, University of Manitoba Airpower Seminar, (November 2003).
30. Bill Keller, "The Fighting Next Time", The New York Times Magazine, March 10, 2002.
31. Allan English, "The Operational Art: Theory, Practice, and Implications for the Future", a monograph written for Canadian Forces College, 19 January 2003, 29.
32. Vincent J. Goulding, Jr., "From Chancellorsville to Kosovo, Forgetting the Art of War", Parameters, (Summer 2000), 8.

"If we do discover a complete theory, it should in time be understandable in broad principle by everyone, not just a few scientists. Then we shall all, philosophers, scientists and just ordinary people, be able to take part in the discussion of why it is that we and the universe exist. If we find the answer to that, it would be the ultimate triumph of human reason - for then we would truly know the mind of God.

Stephen Hawking

THE THEORY OF EVERYTHING

Introduction

We now turn our inquiry to the contemporary state of Western military thought and theory. As we proceed, we should note several tendencies: First, the propensity of Western military culture to look for simplistic if not facile explanations of the complex matrix; second, whether wilfully or by accident, Western society's lack of historical foundations demonstrates a poor appreciation of the present due to misinterpretations or ignorance of the past. Whether there has been a Revolution in Military Affairs (RMA), the validity of Manoeuvre Warfare (MW) theory, Effects Based Operations (EBO), the roles of technology, 4th Generation Warfare (4GW) and whether Military Theory in general is evolving to offer humankind a better understanding of the nature of war will all be discussed. We will investigate the issue of change wrought by new technology and the effect, if any, that it has had upon the nature of war. From the perspective of the model presented in the 'Introduction', technology has immediate effects upon Tactics, Techniques and Procedures but less influence upon any cogent theory. Whether technology has any effect upon philosophy remains unresolved.

Alexander the Great reputedly wept when he saw the breadth of his empire, for he realized that there were no worlds left for him to conquer. Western society may once again be approaching such hubris with the announcement in the popular press that mankind is on the verge of discovering the 'Theory of Everything.' Humankind has been striving to better understand war's true nature for as long as it has been searching to understand the nature of the universe. Military theories are thus analogous to scientific theories; many models and explanations have been proposed in both fields over the millennia. Some have assisted in humanity's understanding; others have been dead ends. Perhaps the value of searching for a theory that explains everything is not in finding the theory but rather in the search itself. This long-awaited theory holds the promise of explaining how the universe is constructed. Beginning with a theory that explained the world in terms of earth, wind, fire and water, humankind has evolved its theoretical understanding of the physical universe to the point where we dare to believe that we may finally have arrived at a point where we can gain a complete understanding of the universe. Armed with this understanding, scientists hope to then gain an accurate understanding of the nature of light, of gravity, and of creation itself. Thereafter, the premise is that light, gravity, and the physical laws of nature can then be better used – if not manipulated. If nothing else, the quest for a single unified theory, whether in war or in science, keeps humanity aware of the horrendous consequences of ignorance.

As new technologies emerge, they frequently affect every day human life; but do they change the nature of that life? Is the effect an improvement? Is the new better than the old? Is a new theory, or a new interpretation of some phenomenon, necessarily better than one that has been extant for some time? Arguably, there might be rare occasions when a technological tool is developed, which does have a profound affect upon the nature of human life. The ability to splice a human genome and the subsequent technology that allowed cloning might fall into this category; but the current debate over human cloning highlights the dilemmas that face humanity when such fundamental choices confront society. Cloning technology may well be able to change the nature of human life. That is a question perhaps best left to ethicists and theologians. For our purposes, the question is different, subtler. Plainly stated: Is it an improvement?

The answer is less important than the question. Asking the question puts us in the correct frame of mind to consider the connection between technology and the nature of war (as opposed to its relation to the nature of warfare). Momentarily accepting that for the most part technological change tends to make humanity more comfortable, safer, freer to pursue leisure, we may wonder if technology does not do the same for war. The quest by military philosophers, theorists, and strategists for a unified theory of war has been analogous to the search by scientists to find one single theory to explain adequately all the universe's behaviours. In a likewise iterative if somewhat less scientific process theories of war have evolved and changed with advances in science, technology, and society. Just as with the advances in the understanding of time and space from scholars like Albert Einstein and Stephen Hawking, so the evolution of war from simple foot soldiers to cavalry and chariots and from muscle to chemical power, has brought the belief that the future of war bears little resemblance to the past. Reminiscent of the predictions that machine guns, air power and nuclear weapons would make soldiers obsolete, 'network-centric warfare,' 'Information Warfare' and '4th Generation Warfare' are all popular concepts in current military circles that once again make such claims.

The correlation between technology and warfare is well-studied and the development of weaponry and technological enhancements to equipment and procedures also well-documented; but is this relationship well understood? Reading surveys like Archer Jones', *The Art of War in the Western World* is informative but leaves the reader wondering whether technological advances alter the nature of war, or merely make the killing more efficient, the manoeuvre more accurate, the fires more deadly.

> One of the questions that is often sidestepped in these discussions is whether advancements in technology can fundamentally change the [nature] of war. Classical theorists suggest that the essential nature of war is immutable ... On the other hand, it is difficult to argue that technology has not been a factor in warfare. In 1298, for example, it was the English use of the longbow that broke the line of the Scots at Falkirk; the same technology was used to similar effect against the French at Crécy in 1346, at Poitiers in 1356, and at Agincourt in 1415. But had technology changed the nature of war? While the French suffered repeated defeats, the Scots learned their lesson at Falkirk, and when they fought the English again, just 16 years later at Bannockburn, they held a contingent of cavalry in reserve to attack the English archers as soon as they appeared. The archers broke and the English were routed. Clearly technology has been able to affect the outcome of individual battles, but can it change the nature of war?[1]

Certainly, technological improvements change the *procedures* used to fight. They alter *warfare*. But does the evolution of technology change the *nature* of war? The evidence so far implies that it does not.

Revolutions in Military Affairs

As with any profession, the military has a community of journals where writers of all stripes can discuss ideas both new and old. Occasionally a concept captures the collective imagination of many of these writers. Possibly the most potent of these notions in the past several decades has been the declaration that the 20th century was marked by an RMA:

> Modern military journals are replete with articles claiming that recent advancements in technology constitute a Revolution in Military Affairs (RMA). The authors of these articles claim that innovations in weapon systems – for example, the development of precision guided munitions – and the capacity to wage network-centric warfare are symptomatic of this RMA, and will afford the United States an unprecedented level of situational awareness and the ability to apply force rapidly, accurately, and precisely without fratricide or collateral civilian casualties.[2]

Robert Bolia cites two pairs of authors: Arthur K. Cebrowski with John J. Gartska, and Bill Owens with Ed Offley. Interestingly, Williamson Murray, in "Thinking About Innovation," *also* cites Owens and Offley – but in a most unfavourable light; Murray accuses of being 'disturbingly ignorant of history' and chides them 'for astonishing misstatements of historical fact.'

In the decade of the 1990s, work that had been ongoing quietly for some time suddenly became widespread. What the US Government's Office of Net Assessment (ONA) had originally called the Military Technical Revolution (MTR) became widely known as the Revolution in Military Affairs (RMA). If any one individual can be credited with lending legitimacy to the RMA it was the ONA's Director, Andrew Marshall. A shy and reticent man, he had a powerful influence upon senior military officers in the Pentagon. Marshall was convinced that "large strategic gain could be achieved through radical advances in military technique based upon the intelligent investment of resources in technical innovation."[3]

Although now faded, it did not take long before the RMA was the favourite hobbyhorse of most military authors, particularly in the US. Each author who espoused the RMA usually explained that it had manifested itself in some singular fashion, or that it had had an impact so fundamental as to change the very nature of war. This manifestation was then interpreted in an equally large and varying number of ways. For example, Jane Wales' "US Nuclear Plan Signals a Policy Revolution" claimed that the use of nuclear weapons altered the nature of war. The United Nations Institute for Disarmament Research talked of the RMA changing the nature of war because of the effects of disarmament. Still others described the RMA in terms of asymmetric warfare. These are a mere sampling of the literally *thousands* of articles and claims on the RMA that followed.

Interestingly, there was one aspect of each version of the RMA that was consistent: the constant claim of a technological leap forward which made mankind's incumbent understanding of war obsolete and therefore called for a new understanding of war. Based on empirical data, these claims usually stemmed from observations of conflicts that followed unpredicted paths. The war in Afghanistan was not what the US Army expected in the same way that it thwarted the Soviet Red Army; the first Gulf War caught the US by surprise at how quickly it gained victory; Vietnam was not the war that the Korean veterans were prepared to fight. The Korean War bore little resemblance to World War Two. The trend was clear.

But upon reflection, the claims did not withstand scrutiny. "There is a perennial temptation to misread recent and contemporary trends in warfare as signals of some momentous, radical shift. As often as not, the character of warfare in a period is shaped, even driven, much more by the political, social, and strategic contexts than it is by changes integral to military science."[4] Conversely, sustained periods of peace may cause military thinkers to lose touch with reality. "A long peace, one that lasts forty or fifty years, could well create military cultures that no longer understand the fundamental nature of war, in which planners assume that there will be little friction or that opponents will be unable to interfere with the conduct of operations."[5] More than any other factor, the development of technology in warfare continues to spur a range of pundits and soldiers to predict some ill-defined but radical shift in war's nature. Perhaps this is because technological change is so immediately tangible; perhaps it is simple wishful thinking; perhaps it is a fundamental misunderstanding of war's nature. Whatever the reason, the claims made by the heralds of technology have invariably been wrong. History is replete with examples of advances in technology promising more than they delivered. To borrow an old American expression: 'a day late and a dollar short.'

As we saw above, Queen's University of Belfast Professor Michael Roberts first coined the term 'The Military Revolution' during an address in 1955. He argued that the warfare during the century 1560-1660 had undergone a rapid and a fundamental change and that this change had been driven primarily by technology. Published the following year as an article, Roberts' thesis has stirred debate in the historical community ever since; and for many years his claim was accepted by historians such as Simon Adams and William McNeill. Some, like Geoffrey Parker, himself a respected historian, took the argument even further. Parker subdivided Roberts' revolution into four parts, which Roberts himself did not do. These divisions highlighted the major implications of the changes wrought during the period: the revolution in tactics; the rapid growth in the size of armies; the adoption of more complex strategies necessary to bring large forces to battle; and the vastly increased cost of warfare for society – not only in financial terms but also in social costs such as depopulation, administration, and damage to infrastructure. It was difficult to find an article or an essay of the period that did not include some mention of this 'revolution'.

Although correct in his claim that technological change transformed the era's armies and navies, Roberts was otherwise mistaken. Real revolutions are not about technology; they are about new ideas. Revolutions occurred whenever:

> ...a new technology was matched with a new strategy to make great leaps of military advantage. Seven centuries ago, archers with six-foot bows rendered heavily armored knights on horseback obsolete. Infantry became the dominant form of warfare, chivalrous rituals of combat gave way to blood-soaked battlefields and, because bowmen were cheaper to train and equip than horsemen encased in armor, minor powers like Flanders could hold their own with great powers like France. At the outset of World War II, the Germans rolled across much of Europe because, between the world wars, they liberated tank units from the slow-marching infantry, provided them with supporting aircraft linked to the ground by the novelty of two-way radio and used these aggressive formations to smash holes through enemy defenses. The *Blitzkrieg*, as this lightning air-and-ground strike was called, replaced a warfare of fortifications and slow-moving foot soldiers with a warfare of maneuver. In that same lull between the great wars, the Americans and Japanese simultaneously refined the aircraft carrier, creating portable islands of power to float where they were needed and transforming naval battles from great slugfests of armored vessels

into a thrilling choreography of aircraft.⁶

It would be easy to mistake the message above. If we were to focus upon the archers instead of upon their use, then it could be incorrectly argued that archery led to the downfall of knights in armour. If we were to focus upon the tanks used by the Wehrmacht instead of the idea that armour should be unleashed from its foot-slogging infantry (as has been done is so much popular literature), then it could be mistakenly argued that the German army was so successful because of its superior tanks.

All these arguments share a common flaw: They focus on the wrong aspect of the cause-and-effect process. The archers defeated the knights because the latter were misemployed; the best tanks in 1940 were French Peugeots, whose superiority was rendered useless by poor doctrine.⁷ If we were to focus on the development of aircraft carriers instead of the concept of extending the striking power of fleets, we could wrongly think that aircraft caused a revolution in naval warfare when what caused the change was understanding how to use aircraft to project power without hulls in the water. Thus, it is more than merely perspective; it is understanding what is being investigated and not confusing the tool with the concept underpinning the tool's use. Sadly, there is a propensity for many military officers to pin their hopes on future weapons systems and slick PowerPoint presentations to solving past problems:

> ...even the most sympathetic onlooker is likely to sense that the Pentagon lives in a sea of slogans, briefings using elaborate electronic graphics, and a self-satisfied belief that new platforms will solve the tactical and operational problems of the future. Unfortunately, slick presentations do not equate to serious military thought. Nor does the procurement of sophisticated – and therefore exceedingly expensive – weapons systems necessarily lead to a "revolution in military affairs." In fact, technology has rarely been more than an enabler of revolutions in military affairs in the past, and there is no reason to believe that things will be different in the future.⁸

Revolutions do not occur because of new technologies; they occur when new ideas harness technology because "no matter how magnificent", technology does not of itself "constitute a revolution. True revolutions happen, above all, in the minds of men."⁹ The Zulu Wars offer an excellent example where technology was possessed by both sides but only the side with appropriate doctrine was able to profit:

> In 1879, in South Africa, at the battle of Rorke's Drift, some 4,000 Zulus were equipped with 800 captured Martini Henry rifles sited at 1,000 yards. The Zulu force surrounded a garrison of only ninety-eight able-bodied British soldiers. However, the Zulu warriors fired their rifles as individuals; they shot high and sporadically and succeeded in killing only sixteen British soldiers. In contrast,

before the battle at Rorke's Drift started, British sergeants were preparing to deliver deadly volley fire in a method of using firearms derived from Greek formation warfare. Through technology combined with technique, the vastly outnumbered British won the subsequent battle.[10]

Although active interest in a so-called RMA seems to have reached its apogee in the last decade, with literally thousands of essays and on-line articles, this interest has really been a renewal of the original awareness, spawned by Roberts, which preceded it by almost a half-century:

> We are at present in the midst of the greatest military revolution of all times. It actually began in the final phase of the Second World War with the explosion of the atom bomb over Hiroshima, but it is only now, in the fifties-and it will be still more so in the sixties ... that we are beginning to understand its real scope.[11]

Whenever such interest originally began, the question of an RMA has not been demonstrated convincingly. In the rush to embrace the alleged RMA, many professional soldiers, as well as many academics have been quick to talk about the changed nature of war. In fact, the RMA has been spoken of so frequently and by so many senior military officers and analysts, that it becomes easy to accept its implicit validity.[12] In the 1930s, the Nazi Ministry of People's Enlightenment and Propaganda ably demonstrated that any assumption, repeated often enough, would be generally accepted as truth. The often highly edited misinformation in mass media today on sites like Breitbart News or even Fox News are more modern examples of the same process.

Similar misinformation regarding warfare is plentiful and widespread: war is more complex than in the past; the battlespace is more crowded; war is more lethal than ever. Consider this last example. The empirical evidence suggests the opposite. Browsing through the casualty lists of any of the wars of the 18th, 19th, 20th and 21st centuries does not bear out the supposition that fighting has become more lethal than it was at Waterloo, Gettysburg, or Normandy. More soldiers died in all those battles than in the entire course of the current wars to liberate Afghanistan and Iraq. In approximate figures, Waterloo saw 83,000 casualties in three days; Gettysburg 51,000 casualties also in three days; Normandy 400,000 casualties over the course of a month-long campaign. Casualties in Afghanistan and Iraq pale by comparison. Without doing a statistical analysis comparing numbers involved, removing air force casualties for modern engagements, comparing killed to wounded and so on, it is safe to say that the modern battlefield is *not* becoming more lethal – quite the contrary.

If the claims of RMA proponents that the nature of war has undergone some fundamental changes are true, why has the proof of any fundamental change not been forthcoming? A basic premise of the revolution has been that technology has made it possible to know almost everything about your opponent. Modern commanders can gather more information about their enemies than ever before, but this does not equate to intelligence. The phenomena of 'information paralysis' and 'paralysis by analysis' would seem to suggest that the trends are moving in the opposite direction. This is due to constantly waiting for that one golden piece of information, which will allow decision-makers to put all the pieces of the puzzle together. Obviously, this golden nugget never does fall into the lap of the commander but in the meantime, the enemy gains the initiative and the headquarters' ever diminishing willingness to accept risk traps subordinates. 'Paralysis by analysis' is a similar process whereby commanders and their staffs endlessly seek the latest piece of the puzzle and are constantly analyzing and reviewing their data in the hope that surety, and thus security, will arise from this analysis. The sad result is that timely decisions are not made, and initiative is lost. Recall General Patton's admonition (not original to him) that a poor solution immediately is better than a perfect solution too late!

Naturally, none of this is new. During the US Civil War, President Lincoln fired at least one commander of the Army of the Potomac because he had 'the slows', that the general was indecisive and insufficiently aggressive. The attack by the US army on Baghdad in 2003 seems a perfect example: Although the Americans had numerical, technical, technological, and information supremacy, their forces still managed to blunder into ambushes just as would have been seen in the Second World War.

Surely, to determine whether the nature of war has changed, we must first know the nature of war. To prove that something has changed, we must first be able to describe what it once was. Admittedly, modern weapons fire longer distances. They strike with greater accuracy. Computers have enhanced an armed force's ability to send data faster and farther. Aircraft are faster and munitions are 'smarter'. Microchips or processors, which them to perform semi-autonomous functions like distinguishing friend or foe or selecting a target from an array. Some of the effects of these processes can be quite astounding. A single soldier on the battlefield can now 'paint' a target with a laser and an aircraft flying ten miles above him can release a weapon, which will acquire the painted target and then strike it. The French army main battle tank *Leclerc*, for instance, has an automated computer aided target recognition system that will 'see' an enemy vehicle, aim at it, load the optimal ammunition to kill it and then ask the commander for permission to fire. Microchip technology has permeated every aspect of military life from ration accounting to target acquisition. Time and space have taken on new shades of meaning. But what does all this mean as a reflection of the nature of war? How can we say that the nature of war has changed?

The truth is that the nature of war was never really understood in the first place. Certainly, some of the claims, from time to time, may have been justified. Occasionally, some technological or sociological aspect of warfare may change in such a fundamental manner as to merit being called revolutionary. The change from muscle power to chemical power, – from horses to internal combustion engines – might be argued as one such event. Yet this claim begins to weaken when we consider that mechanized armies still

THE THEORY OF EVERYTHING

took to the field to fight each other – they just had machines to ride instead of cavalry mounts. We might argue that converting warships from sail to combustion engines was such a fundamental change. But modern navies still 'sailed' out to sea after the last sail was thrown away and replaced with coal, diesel, and nuclear power – they just had to consider the need to reprovision their ships instead of only their men.

Determining whether there has been a revolution is not so simple. After all, life itself is change. Within this change, some aspects remain constant while others slowly metamorphose into new forms. The difficulty lies in discerning whether the change, which is being investigated is *evolutionary* or *revolutionary* in nature. For example:

> [T]he impression historians form can depend on the cases they select and the contemporary sources they consult. The devastating victory of German forces in the campaign against France in 1940 would seem as clear a "revolution in military affairs" as any in the 20th century. Yet virtually none of the German generals responsible felt there was anything revolutionary in that victory. In fact, one of the most perceptive General Staff officers, General Erich Marcks – soon to be selected by the army's chief of staff, Franz Halder, to draw up the initial plan for the invasion of the Soviet Union—noted in his diary in late June 1940 as the major explanation of the success in France the ideological motivation of German soldiers. On the other side of the hill, however, his counterparts in the British and French armies clearly believed that something revolutionary had occurred.[13]

The trend during the last few decades has been for heralds to declare that everything has changed in a revolutionary way each time a new technological advancement appears. Recall Alvin and Heidi Toffler who argued in their latest work that warfare had undergone revolutions only twice before. They used the term 'wave'; The 'First Wave' was when society changed from an agrarian base to an industrial one. The 'Second Wave' was when society modernized. According to them we are now in the 'Third Wave' as warfare is transformed once again to post-modern or information-based warfare.

But Alvin and Heidi were not alone. A separate call to re-invent war occurred in the USMC in-house *Gazette*. In the late 1980s a group of military thinkers declared that the current 'third generation' of warfare – not to be confused with the waves – was obsolete. According to their argument, we had entered a 4th generation of warfare (4GW). Like a star becoming a nova, the idea flared, gained wide acceptance, and then perished. It did not enter the doctrine of any army in a substantial way. But ideas like EBO, Information Warfare, and Cyberwar rise and fall frequently. Doctrine writers and commanders in the field, must be vigilant and have the intellectual skills that allow them to distinguish an authentic change from a simple fad.

The advocates of technology are not uniform in their visions of where warfare is headed. The future is an untapped resource and opinions, both informed and otherwise,

> **GENERATIONS OF WARFARE**
>
> **1st Generation:** Age of Napoleon
> **2nd Generation:** Age of Firepower
> **3rd: Generation:** Age of Manoeuvre and Ideas
> **4th Generation:** Age of Asymmetric Warfare

abound. Although many of these opinions share common elements, they are by no means homogeneous:

> [Franklin C.] Spinney and other Boyd apostles, who sometimes label their work 4GW, shorthand for fourth-generation warfare, have many differences with the RMA camp, but their main complaint is that Marshall and his followers place too much faith in technology, promoting it as a way of removing discretion from the battlefield and having the battle managed from a distance, by 'generals in chateaus,' as Spinney likes to say. Spinney is, of course, dubious of the claims for high-tech weapons in Afghanistan.[14]

Heterogeneity does not detract from the common belief that war has undergone some type of fundamental change and that the 'old' rules no longer apply as they once did.

A good case study concerning the notion of technology altering the nature of war is the concept of Information Warfare. By looking at a possible RMA through the lens of Information Warfare, we can investigate the claims made by all proponents of RMA in whatever form they may be championing their cause. Of the various manifestations of RMA, most have a basis in the concept that information is the most potent weapon of post-modern societies and that the critical vulnerabilities of all fighting forces are their command, control, computer, and information (C3I) systems. The resulting premise at its heart is straightforward: destroy the C3I of your enemy and they cannot fight. The commander who could achieve this objective could hope to gain Sun Tzu's ultimate victory; the triumph won without battle.

This is not a new idea. What is new, however, is the concept of "information superiority" – sometimes called "information dominance" or "information operations," – a revelation that seems to have come relatively recently. Once again led by the US, authors began to speak almost in euphoric terms about how information-based societies would fight wars in a fundamentally different way than their industrial-based precursors. Subsequently, professional journals began to fill with essays on information warfare. The

THE THEORY OF EVERYTHING

	1st GENERATION	2nd GENERATION	3rd GENERATION	4th GENERATION
NATURE OF THE PROBLEM	State Monopoly Disorderly Mob	Mass Armies Attrition	Blitzkrieg Dislocation Non-Linear	State Losses Monopoly on War 3 Block War Non-State Actors
NATURE OF THE SOLUTION	Line & Column Culture of Order	Mass Firepower Synchronization Discipline Conventions	Auftragstaktik Decentralization	Peacekeeping Peace Support Nation Building ISTAR

predictions of a fundamental change in war's nature came quickly with most authors offering concordant views of how the old understanding of war was simply not valid in the modern context. But despite some dire predictions, not all military thinkers were convinced:

> Current interest in information warfare and the manifold effects of the information revolution on the conduct of war cause many to proclaim a revolution in warfare. Evangelists of information warfare, like forerunner evangelists of air power, sea power, and artillery, risk losing sight of historical context and the continuities of conflict. We are once again faced with a genuine technological revolution which seems to offer an entirely new mode of warfare, one that advocates insist will supplant existing modes.[15]

Such predictions can have a powerful pull upon the mind. They enrapture their audience with the unspoken promise of a 'bloodless war.'

Like information warfare, the concept of bloodless war is not new. Clausewitz discusses it in his writing and warns the reader not to believe that such a thing exists. A corollary of bloodless war is the already mentioned aversion to casualties. Certainly no one would wish to see needless casualties nor a return to the bloodletting that was seen in the two world wars. But particularly in Western liberal democracies, where armies consist of citizen-soldiers, at times public aversion to casualties has become completely unrealistic. The Vietnam War was an early example but even in the Gulf Wars, and later in Afghanistan, it was common to hear the claim that the high casualties were unacceptable. But what might be acceptable? A military force of several hundred thousand US soldiers projecting power 10,000 miles from their homeland that suffers two or three soldiers killed per day, however tragic, does not seem unduly high, especially when viewed historically. A survey of back issues of *The Stars and Stripes*, the US Armed

Forces' newspaper, shows that on average the US military loses approximately the same number of soldiers every day just in peacetime training accidents. This issue of casualty aversion is a direct consequence of the vision of a bloodless war that the proponents of these new forms of war worked so hard to sell for so long.

The issue of casualty-free war is a key component of the belief in a technology-driven RMA and that we need merely allow this technology to be properly employed. It is the air power argument in a different garb and in this context, it is easy to understand the temptation to declare that a revolution is at hand. But we must distinguish between superficial declarations and manifestations and the underlying fundamental factors, which govern all war. Theory, however elegant, must always be tested in the laboratory of battle:

> Knowledge-based warfare sounds good in theory. It creates amplitudes of military effectiveness. The reach that technology and intelligence and knowledge management give us, allow for death to be administered at a distance, on a large scale, where the messy details do not impinge. Even our vocabulary disengages our emotions and allows us to pretend that this is a clean, smart, knowledge-enhanced project. And the precision of our munitions allows for the most part, our claim that we can discriminate between civilian and military casualties.
>
> But if the enemy suffers from knowledge-enhanced warfare, the sword is double-edged in unanticipated ways. In the first week of the Iraq invasion, friendly fire and accidents accounted for two-thirds of coalition deaths. Our complex military machines and our enhanced intelligence capabilities are sometimes more lethal than even our foes.[16]

For a short time, the argument that modern technology had changed the nature of war gave rise to the legitimate questioning of whether the 'timeless' principles of war remained valid. There was considerable debate, about whether the principles of war had any value in the Information Age. The debate was not new: as early as 1966, in a lecture given at the US Air Force Academy, Professor Peter Paret was convinced that the principles were not so much an intellectual tool as they were a crutch. He referred to them as:

> A catalogue of commonplaces that since the beginning of the 19th century has served generations of soldiers as an excuse not to think matters through for themselves. In Napoleon's time, the principle of concentration of force made operational sense, especially when it was brought about by high mobility, separate advances, and the indirect approach. When in his later years Napoleon tried to apply this same principle to tactics, pressing his infantry into solid, ponderous masses, whose path was to be cleared by a vast accumulation of artillery, the strategic concept degenerated into a self – defeating tactical

THE THEORY OF EVERYTHING

absurdity. Its validity in the mid – 20th century remains at least in doubt.[17]

Unfortunately, like most of the later critics, Paret missed the point. He confused the tool with the artisan: was the table badly built because of poor workmanship or were the tools inappropriate to the task? Look, for instance, at his example of the principle of concentration. It was certainly true that when Napoleon pressed his infantry into immobile masses, they became tactical liabilities. Nonetheless, consider the demonstration of the principle of concentration during the 1944 Normandy landings:

> June 6, 1944, witnessed the largest concentration of air, land, and sea forces in the history of the world. On, over, and along roughly 50 miles of the French coast were eight divisions of Allied ground and airborne soldiers, 5,000 ships, and 7,000 aircraft. Never in war had so much been concentrated at one point, at one time. The German Luftwaffe had been designed for just such a contingency – a tactical air force created to support the German army. So concentrated were Allied forces that conceivably any German bullet fired, any bomb dropped, would find a target of some kind.[18]

Again, in the 1990s – more than two decades after Paret's speech – by the massing of air power against the Iraqis during the Gulf War:

> The air campaign involved nearly every type of fixed-wing aircraft in the U.S. inventory, flying about 40,000 air-to-ground and 50,000 support sorties. Approximately 1,600 U.S. combat aircraft were deployed by the end of the war. By historical standards, the intensity of the air campaign was substantial. The U.S. bomb tonnage dropped per day was equivalent to 85 percent of the average daily bomb tonnage dropped by the United States on Germany and Japan during the course of World War II.[19]

The validity of the principle, then as now, continues to be moot; but the real question remains whether the principle's application correctly considers time, space, and circumstance. I am not arguing for or against the principles of war. My digression is merely an example of how simple it is to make a bald statement regarding the need to sweep away the past in favour of some newly uncovered rule, principle, or truth. In the absence of proof to the contrary and considering the historical success that so many commanders have claimed because of adherence to principles – whatever they were – it is safe to leave sleeping dogs lie and side with Fuller and NATO doctrine generally. Principles live on, despite information warfare.

In the end, the hyperbole regarding information warfare, information dominance and information operations has faded. More reasonable voices have expressed the belief that war has not really changed. Rather, information became an adjunct to the other weapons and the doctrine on how to integrate information operations within the battle rhythm of a military operations continues to develop. There is less talk of the exclusivity of information operations at the expense of all others. In fact, there has been a resurgence of military professionals considering what many tried and true classical theorists have said for many generations. Whether we return to Sun Tzu, Machiavelli, or Clausewitz, the message is similar and familiar. War is a human enterprise:

> Clausewitz reminds us that the human elements of war are extraordinarily difficult to gauge. If necessity is the mother of invention, asymmetric tactics, strategy, or technological countermeasures will always upset the best laid technology-based plans. Nuclear weapons were supposed to make conventional arms obsolete and totally revolutionize warfare. Massive retaliation, the Eisenhower-era strategy for their deployment, ignored conventional capabilities only to find that the will to use weapons of mass destruction was lacking. Dirty little wars on the periphery continued and guerrillas flourished, despite our fearsome nuclear arsenal and the threat of certain annihilation we wielded. Our bluff was called in Korea and later in Vietnam, and the nukes remained holstered. Soldiers at the dusk of the industrial age faced the same mud and mayhem that had confronted their counterparts during the Napoleonic wars at its dawn.[20]

Yet, the return to classicism is not yet complete. In discussing the war in Iraq in 2005, retired Major General Robert Scales, a past Commandant of the US Army War College – where future generals are sent to shape future US doctrine – echoes the frustration of many who have tried to impart to younger leaders of the US armed forces the benefit of their experience:

> Intimate knowledge of the enemy's motivation, intent, will, tactical method and cultural environment will prove to be far more important for success in the advisory phase than smart bombs, aircraft and expansive bandwidth. A successful advisory effort depends on the ability to think and adapt faster than the enemy. Soldiers must be prepared to thrive in an environment of uncertainty, ambiguity and unfamiliar cultural circumstances. This war will be won by fostering personal relationships, leveraging non-military advantages, reading intentions, building trust, converting opinions and managing perceptions, all tasks that demand an exceptional ability to understand people, their culture and their motivations.

> Yet even after nearly three years of evidence to the contrary the Department of Defense still pins its efforts to fight this war in large measure on the concept of "net-centric warfare." Military theorists in the Pentagon claim that new

THE THEORY OF EVERYTHING

information and computing technologies will allow U.S. military forces to "lift the fog of war." According to this view, a vast array of sensors and computers, tied together, can work symbiotically to see, and comprehend the entire battle space and remove ambiguity, uncertainty, and contradiction from the military equation, or at least reduce these factors to manageable and controllable levels.[21]

Information warfare is not a bad idea. Using technology is and always has been an integral aspect of war and warfare, whatever the type of combat. But what Scales is decrying is the US military culture's increasing desire to become wedded to technology for its own sake, rather than for the outcomes it can create, principally, the saving of lives. He is arguing for what I call 'appropriate technology'.

Scales correctly argues that technologies give military leaders an unprecedented overview of the battlespace. But focusing on position, or as we have come to call it 'situational awareness' is to emphasize the wrong aspect of our information dominance. In the end, the analogy of the chessboard is apt. During a game of chess both players can see all the pieces that both sides have at their disposal. This does nothing to guarantee victory for either player. What cannot be seen is the intent that each opponent has; what cannot be measured is the manipulative skill that each player has in combining and recombining his pieces and manoeuvres; what cannot be seen is the will to overcome, adapt, dominate, and win.

Manoeuvre Warfare (MW) Theory

Just as information warfare is a good case study to demonstrate the logical fallacy of technology altering the nature of war, MW shows how change occurs as an outcome of ideas, and how a refreshed theoretical understanding of the nature of war (even if it is not an improved understanding) can change warfare dramatically. In the early 1980s, American congressional aide and amateur military theorist William S. Lind published a booklet entitled *The Maneuver Warfare Handbook*. Lind had formerly published an article "Some Doctrinal Questions for the United States Army," in *Military Review* in March 1977 in which he coined the somewhat misleading term Maneuver Warfare.[1]

Lind described a type of warfare that had been discussed for several years. It was warfare reminiscent of the lightning and lopsided victories of the German Wehrmacht's invasion of France in 1940 and of the Israeli Defence Forces' striking victories during the 1967 Six Day War and the 1973 Yom Kippur War. More importantly, Lind described a style of warfare that emphasized the human aspects and the chaotic nature of war. He dubbed his fusion of old ideas Maneuver Warfare because his synthesis stressed the power of movement and of action both physical and mental; it emphasized working within chaos; it underlined strong leadership at all levels. Most importantly, it subtly but strongly downplayed the technological factors of warfare while exalting the human

---

1 Maneuver and Manoeuvre are US and Canadian spellings.

ones. In an almost heretical break with accepted wisdom, Lind's book told battlefield commanders to disregard ground as a dominating factor in victory and to focus instead on defeating the enemy's morale. Although originally ridiculed as foolishly simplistic, MW quickly captured the imaginations of senior US Marine Corps leaders. Not long thereafter, US Army leaders came to incorporate the tenets of this theory into their doctrine, albeit with modifications that caused some observers (like me) serious pause.

An example illustrates the point. MW encouraged commanders to build synergy with their subordinates, peers, and superiors through *Auftragstaktik* and initiative acting within the principle of unity of effort. The USMC applies this by training commanders to think independently within a coordinated whole. The US army did the opposite. They preferred the use of a staff tool called a 'synchronization matrix.' MW called for synchronization to be an outcome, which the USMC seemed to understand. The US army called for synchronization to be a tool or process – a fundamentally different understanding of the theory. The British army, following neither the Marines nor the US army, molded the theory into a third and completely different process.

In 1989, I introduced this concept to the Canadian Armed Forces in an article, "Smaller Can Be Better", published in the *Canadian Defence Quarterly*. The idea was widely panned as unworkable but once the Americans adopted MW, I found earlier opponents evangelizing the idea to me as the latest invention that would revolutionize warfare. Although widely and differently interpreted and not yet universally understood, MW soon found its way into all the major land forces of the NATO alliance and is now the basis for the military doctrine of almost all NATO armed forces.

The US, Great Britain, Canada, and NATO as an alliance have all rewritten their doctrines to incorporate aspects of MW theory. Unfortunately, no single theory is as poorly understood as MW. Just like Clausewitz, the language of MW is universally quoted by NATO officers and ironically, after four decades of open debate and doctrinal study, the concept is *still* not well understood, but that does not stop it from forming the cornerstone of most NATO fighting doctrine.

Contrary to common but mistaken belief, MW has nothing to do with manoeuvre *per se*, which is widely defined as the combination of fire and movement; rather, MW is an overall *style of warfare*. This style stresses rapid decision-making and independent action, which takes advantage of battle's inherent chaos and Clausewitzian *Friktion*. By stressing rapid decision-making and low-level initiative, commanders and subordinates at all levels are taught to *dislocate, disrupt,* and *disorient* their opponents, *destroying their cohesion*, rather than destroying them physically, piece-by-piece, with firepower. In other words, make the strength of the opponent irrelevant while applying one's strength to dislocate, disrupt and disorient him. In MW, a commander seeks to create *successive, unexpected,* and *threatening* situations for his opponent. The opponent should be brought to see his situation not just as unfavourable or deteriorating; he must see his own situation as deteriorating *at an ever-increasing pace*. The desired outcome in an opponent is confusion, disorientation, irrelevant decision-making, and mental paralysis leading to

THE THEORY OF EVERYTHING

his eventual defeat.

The term 'Maneuver Warfare' has proven to be an unfortunate misnomer for this style of warfare. Although making use of manoeuvre, it is more about how a commander interprets battlespace and then applies combat power to achieve a desired outcome. Lind had been inspired by the sweeping armoured manoeuvres of commanders like General Heinz Guderian, Field Marshal Erich von Manstein and General Moshe Dayan. Choosing the term to convey a flexibility of thought and action combined with a rapidity of tempo, Lind redefined the rate at which a commander, or organization, can move from stimulus to response and then return to the beginning of the cycle. In general, fire was to be used *to create favourable conditions for manoeuvre* and to *annihilate units whose cohesion has been shattered*, not to engage cohesive enemy units head-on in an attrition contest.[22]

Although MW was not new *per se*, it was new conceptually as a style of warfare. Lind chose the conflict theories developed by Colonel John Boyd, United States Air Force as his intellectual foundation. Boyd, a self-educated fighter pilot, who had a strong influence on the desig of the F-16 fighter, had become infamous within the inner circles of the USAF for his controversial theories on aerial combat. Curious to explain the categorical casualty disparity between US pilots and their Communist enemies in aerial dogfights during the Korean War, he went to the archives to study the combat data. He wondered why American pilots had a kill ratio of better than 10:1 over their North Korean and Chinese counterparts. In many respects, the Russian designed and Chinese-built MiG fighters were better than American aircraft and certainly, the pilots were not so dissimilar in training and technical ability as to justify such a ratio. Boyd analysed the data using his own self-taught methods and soon had his answer. He posited that every conflict consisted of a series of time competitive-observation-oriented-decision/action cycles. This cyclical series soon became the eponymous Boyd Loop or the now more generic Observation Orientation Decision Action Loop or merely the OODA Loop.

What Boyd discerned from his investigation was that Allied fighter pilots were able to observe their targets, orient their aircraft relative to their enemies, decide on a course of action and implement that decision more quickly than the Chinese and North Koreans, who were controlled by ground-based aerospace controllers and not taught to think or act independently. As the adversaries chased each other around the sky, the Allied pilots cycled through their OODA loops so quickly as to make their enemy's actions not just irrelevant but *increasingly* irrelevant. The ultimate result was that either the Communists were shot down or they would succumb to mental paralysis, causing them to crash.

Boyd then generalized his evaluation: a commander assessed a situation from his observations, decided, and then implemented this decision either through direct action or through his subordinates. Having gone through the sequence, a good commander would begin immediately anew with observation and assessment. He would continue this process until either he gained a victory or lost the ability to work through the sequence. Lind took this cycle and incorporated it into his handbook. MW was the

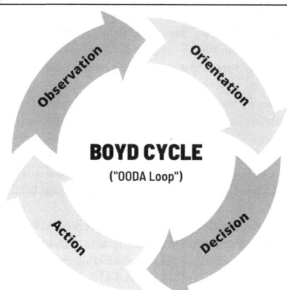

eventual product of his Hegelian process of thesis, antithesis, and synthesis.

Boyd's theory of conflict was Clausewitzian but predicated on the concept of tempo. In other words, the basis of his thinking was philosophical, emphasizing mental prowess and non-physical components of combat. The key for him was winners not only out-thought their opponents, they also did it faster and thereby controlled the tempo of the fight. He postulated that if one commander could push himself through the OODA loop faster than his opponent, he would gain a temporal advantage. Each time he beat his enemy through the loop, his advantage would increase because his opponent would be reacting to situations and actions, which had already lost their validity. More and more, the slower and less mentally agile commander would be reacting to situations which *were* valid but which the more mentally agile commander had changed too quickly or too radically for the slower commander to cope. The inevitable result of this inability to handle a continuously deteriorating situation would be mental confusion and disorientation, as the losing commander more and more found that he could not deal with the constantly and radically changing situation. In the end, the slower commander would either make a tactical error in judgement, which would lead to his destruction, or he could not grasp the meaning of the rapidly altering situation and would suffer from mental paralysis leading to his surrender. Either way, defeat would come to the slower, or less agile of the two commanders. Looking through the lens of Boyd's theory, Lind raised the spectre that perhaps a key element to the battlefield genius of past commanders like Alexander, Gustavus Adolphus, Napoleon or Guderian was their ability to push their way through their OODA loops faster than their opponents. In other words, perhaps battlefield genius was a factor of mental acuity – at the very least it was an intriguing prospect.

THE THEORY OF EVERYTHING

Although the US Marines accepted Boyd's theory as it related to ground combat, where large formations were the norm, armies were more resistant. MW soon became a hotly debated issue in most military journals, but especially in the US military. The sticking points were manifold, but the greatest issue was whether Boyd's tempo/decision studies could be applied to large ground forces. Boyd had been a fighter pilot, who had based his theories of conflict upon observations of aerial combat, where commanders most often met each other in single combat. Due to the nature of this fighting, pilots facing each other could work their way through their OODA loops independently and in relatively short time spans. There was, of course, a decision/action cycle for ground units of all sizes, but due to the numbers of troops involved, and the way in which information was processed through commanders and staffs at all levels, the cycles were necessarily much more complex as well as much longer.

However, the critical point, was not whether the combat was occurring in the air or on the ground. Combat was about an individual, alone or leading a group, making decisions, and acting upon them and, contrary to what is implied in some theories of conflict, the most frequent cause of defeat in war is not physical but rather mental or psychological. USMC Commandant, General A.A. Vandegrift, a man who studied war extensively, and who was himself a battle veteran and Medal of Honor recipient, put it most succinctly in his 1944 book *Battle Doctrine for Front Line Leaders*: "Positions are seldom lost because they have been physically destroyed, but almost invariably because the leader has decided *in his own mind* that the position cannot be held."[23] In the end it is only one person who will decide whether to go on or to quit: the commander.

Was Boyd's theory still valid for non-aerial warfare? Could it be transformed into institutionalized models? The answer was positive. Although not identical to the fighter pilot, the ground commander is nonetheless just as bound to go through the OODA Loop, decide upon a course of action, and then act upon his decision. It is a question of how quickly he, his fellow commanders, and their staffs at all levels can act.

But even before fully embracing MW, the US army (followed closely by the Canadian army) adopted a new command and staff procedure known as the Operations Planning Process (OPP). This process was in effect an OODA loop for staffs, although not described as such, and can be defined much like what it replaced: Battle Procedure, which was the process by which a commander, with his staff, analyzed a situation, decided on a plan, issued orders, controlled its execution, and prepared for further contingencies. The OPP has had a profound effect upon current warfare as Staff Colleges in NATO, especially Great Britain, the United States and Canada, stress this tool and its use (or a version thereof). This is not to say that there has been a shift in war's nature – only that how it is perceived has undergone a paradigm shift. As the Canadian, General Mike Ward, outlines below, staffs have become consumed with the *process* as opposed to understanding the *product*:

Command and staff procedures are focussed on application of the OPP. Broadly speaking, all are adept at using the OPP by virtue of how we have taught and practiced it at CLFCSC [Canadian Land Force Command and Staff College]. As observed in a recent study of UK C2 in OP TELIC however we may have become so focussed on perfecting the OPP products that we have lost sight of the fact that it is a process that aids in decision making and must be applied appropriately in operational conditions in order to enable execution of ops. It must aid communication, brevity, understanding and be timely so as to permit sufficient time for planning and execution at subordinate levels.[24]

Lieutenant Colonel James Storr conducted the study to which General Ward refers. In *The Command of British Land Forces in Iraq, March to May 2003*,[25] Storr made some startling observations: Staffs at all levels had suffered from bloat; divisional staffs habitually sent orders to their brigades *after* the brigades had already begun their combat action; officers on staff were obsessed with form over function. The warnings that had been raised in various journals over the years about officers losing sight of reality, putting too much store in technology, and lacking a fundamental understanding of the nature of war seem to have manifested themselves in real operations. Commanders and staffs that were convinced of a fundamental change in war – because an RMA occurred – were rudely awakened to the fact that the tools may have been new, but the job remained the same.

Misunderstanding Charles Darwin

If doubt has been cast upon the whole concept of RMA, a logical progression might be to wonder if Military Theory has evolved over the centuries to ever-higher stages of development? Certainly, the common impression amongst military officers is that it has, although if writer Stephen J. Gould was correct, then the answer must be that it has not. In his book *Full House: The Spread of Excellence from Plato to Darwin*, Gould contended that Darwin's Theory of Evolution has been misinterpreted to imply that evolution produces an inexorable qualitative improvement in any species. Gould pointed out that what Darwin's *Origin of the Species* really said was that life evolved over time to adapt to its environment and to improve its chances of survival. Darwin did *not* say that this improvement or *evolution* made the species *better*, only that it improved the species' chances of survival.

Clearly there have been changes over time; those living in the present look to the past and are assured by the empirical evidence that humanity is the product of an ever-improving condition – by whatever measure. Military theorists are no different. Each new theory (or hypothesis) begins with the premise that the past has been wiped away by something new and fresh, and *better*. It is taken for granted that modern Military Theory is superior to any Military Theory of the past. Even where older theories, like those of Clausewitz or Jomini, are cited, it is a given that today's interpretation is better

THE THEORY OF EVERYTHING

than that of the past since today is informed by the past. The current doctrine of MW, for instance, is highly disdainful of all military theories that have come before, implying that MW's opposite, so-called 'Attrition Warfare' was the choice of unthinking dullards. Fourth Generation Warfare is likewise so. This is a clear manifestation of Gould's thesis that the common belief is that and natural selection have improved, in this case, Military Theory. But this claimed improvement is an illusion.

The simplest way to attack this question is in reverse order. Let us briefly investigate the current state of a Military Theory that is in vogue among the major NATO partners and look backward to see if there is demonstrable proof of improvement. Before doing so, however, what is meant by 'improvement' must be clear. Better technology is not proof of a better theory; it is simply an increase in technical or technological ability. The fact that artillery can now strike targets over thirty kilometres distant does not prove that current Military Theory is better than that which used artillery that was useful to only 800 yards. This is merely a technical improvement in range. The vast array of telecommunications improvements provides a marvellous increase in the swiftness of passage of information but, again, this increase is one of magnitude and speed and not a demonstrated proof that modern Military Theory is better than what, for instance, came before the invention of the wireless radio. Admittedly, a certain amount of subjective assessment is involved. Nonetheless, remembering the definition of theory – a system of ideas explaining something usually based on general principles independent of what is being explained – it should be possible to determine, generally, if a modern Military Theory offers a better explanation of the nature of war than an older theory.

MW Theory (often confusingly called Manoeuvre Theory, which correctly refers to combined arms combat as developed by the German army during the inter-war period) is neither universally accepted nor uniform in its interpretations. Although all aspects of MW Theory stem from the work of John Boyd, different armed forces have their own schools. For example, the British army, which fully espouses Manoeuvre Warfare, bases its doctrine on a blending of German doctrine (from HDV 100/100 *Allgemeine Führungsgrundsätze*), the writings of Lind and the writings of Brigadier Simpkin (*Race to the Swift*). The German army also espouses Manoeuvre Warfare but has not changed any of its fundamental doctrine since it was rewritten for modern warfare by Hans von Seeckt after the First World War. The American army claims to fully espouse Manoeuvre Warfare and leans towards the writings of Robert Leonhard with, naturally, modifications "bolted" on. The USMC also fully espouses Manoeuvre Warfare but denies most of what Leonhard preaches and falls back on the writings of Boyd and Lind. The US Navy also espouses Manoeuvre Warfare, but of course, their interpretations are informed by the writings of Mahan. So, although all these NATO armed forces agree that MW Theory is the basis for their doctrine, they do not agree on some fundamental issues. The best analogy might be Christianity. They are ecumenical rather than catholic.

With so many interpretations of MW Theory you might be asking yourself what distinguished military fighting doctrines before MW. The answer is 'schools of war' as in

STRATEGIA

the 'French School' – Napoleon and Jomini – or the 'American Way of War' – as already described by Professor Russell Weigley. With the advent of MW, all previous schools have come pejoratively to be known collectively as 'Attrition Warfare' which encompasses all that was bad and brutal about warfare. It has been epitomized by the bloody battles of the First World War, where tens of thousands of men gave their lives over the possession of a few acres of farmland covered with barbed wire. Champions of MW have claimed that practitioners of Attrition Warfare were unthinking clods who simply fed men and materiel into the maw of death hoping to bleed their opponents into surrender. General Sir Douglas Haig, the decorated but somewhat dim Commander-in-Chief of the British Expeditionary Force, and proponent of 'One Big Push,' was a 'high priest' of this style of warfare. Short sharp victories such as the German invasion of France in 1940 and the American victory in the First Gulf War are held up as sterling examples of MW.

Unfortunately for those who believe that MW is an evolutionary refinement, this belief does not hold up under scrutiny. Close investigation of historical examples – if used correctly, and not merely 'cherry-picked' to support one side of the argument – clearly demonstrate that the art of war is no more refined today than it was when Frederick the Great surprised his enemies at Leuthen, or when Caesar defeated Pompey at Pharsalus. As already conceded, technical and technological improvements will make it appear that there may be improvements when, in fact, there are none. This point is well made in the collection of essays on the subject, *The Maneuver Warfare Anthology*, where several authors examine this style of warfare under the microscope of historical fact and conclude that it is neither new nor better than what preceded it.

Current military thought and theory and any concomitant convictions about an RMA have been built, in part, on technology like the ability of electronic jammers to listen to or completely disrupt the messages of an enemy or the ability of supersonic aircraft to deliver bombs to an enemy thousands of miles distant at twice the speed of sound. But how has the greater realm of current Military Theory improved humanity's collective understanding of the nature of war? The answer is that it has not done so in any substantial way. Despite breathtaking advances in technology, we know little more about the nature of war today than we ever did. We may have come up with new paradigms and with new terms, even new concepts but we still return to the ancients for their understanding of the nature of war. We still use the definitions given to us by an ancient Chinese, a medieval Florentine and a 19th century Prussian! This does not say much for humankind's increased understanding of war, and it bodes ill for the evolutionary trend towards ever-higher stages of development.

In many ways, our newest theories of war have moved us further into the shadows than they have towards the light. The current but fading belief that we are amid an RMA is an abrogation of the responsibility to attempt to understand the nature of war by implying that it is changing so fast that whatever we knew before is not now worth knowing. Throwing out the principles of war is a good case in point. Although many do not necessarily place great store in the principles as handed down by God to Fuller and thence to the British army, it is a fair criticism of those authors who are constantly revising the principles, whichever set they may choose, that they have completely misunderstood what these principles represent and that, rather than attempt to gain that understanding, they have simply created newer more alluring principles. This is not refinement; it is a move backward.

The belief in MW does not mean to imply that there has been no progress at all in the development of Military Theory. But just as Gould points out that the so-called refinement of the horse from a three-toed dog to a one-toed thoroughbred has not been one long progression towards excellence, so has it been with Military Theory. There have been individuals who have moved the yardsticks. In the modern era, Machiavelli gave us fresh insights into the nature of conflict, the composition of armies and why and how armies should fight. But these ideas did not spring from his mind like the Titans from the forehead of Zeus. Like any Renaissance man of letters, Machiavelli was well-versed in the ancients. He had studied Vegetius and was interpreting for a contemporary audience. Clausewitz no doubt puzzled his fellow Prussians, not to mention his pupils at the *Kriegsakademie*, by insisting on the dual nature of war. War, he said, had a physical and a moral component. But was this a flash of brilliance or the Hegelian expression of the Taoist philosophy expressed by Sun Tzu 2,200 years earlier when he said that there is a dichotomy in war: that war had a yin and a yang? Force, said Sun Tzu, had a dual component and ever since he said it, Clausewitz, Simpkin, Leonhard and more have repeated it. Even the father of modern armoured warfare, JFC Fuller, in his exhortations for 'brain warfare' over 'body warfare' developed a modern manifestation of an ancient idea: the Arab proverb about a preference for an army of lambs led by a lion than vice versa is reputedly older than Christ.

Summary and Conclusions

Technology, new or old, is not a solution; it is a tool. Recombinant solutions to problems, whether they include new ideas or recycled ones, are what move mankind forward. Although there may appear to be something in human nature that causes many to believe that around the next corner is some technological marvel, which will solve the problem at hand, this is rarely the case. The German popular belief in 1945 that the *Führer* had some secret weapon that would win the war died hard and was clung to by portions of the German population even as the Russians advanced on central Berlin, street by street. The current conviction that some new miracle medicine will finally eradicate cancer keeps many from facing a plain reality that a healthy diet combined with regular exercise is the most reliable prescription that doctors hand out. Regrettably, there seems to be a tendency to attribute more value to something new versus something old – even if it has only the appearance of being new. Modern language is full of subtle reminders not to be 'outdated' or 'old fashioned.' The eminent Oxford historian Cyril Falls expressed this thought – as well as the warning to beware of the apostles declaring an RMA – more than six decades ago:

> Observers constantly describe the warfare of their own age as marking a revolutionary breach in the normal progress of methods of warfare. Their selection of their own age ought to put readers and listeners on their guard. Careful examination shows that, historically speaking, the transformations of war are not commonly violent ... It is a fallacy, due to ignorance of technical and tactical military history, to suppose that methods of warfare have not made continuous and, on the whole, fairly even progress.[26]

Falls is correct; declarations of an RMA are premature because the evidence does not support the assertion. In fact, there is abundant evidence to the contrary; that not only are we not undergoing an RMA now, but that we have never undergone one. Still, technology continues to have a strong attraction as a panacea that will deliver us from whatever dilemma in which we find ourselves. Beginning with Newton and the Enlightenment, with mixed effects, science and technology have been touted as the way of the future.

Aerial warfare offers an excellent object lesson: Supposedly, the reason that aerial bombardment did not win the First World War by terrorizing civilian populations was not that such a strategy was inappropriate; it was because the accuracy and means of delivery was not up to the task. The same arguments were used again after 1939-1945. Strangely, when aerial delivery technology finally was up to the task at the end of the century, it still did not win wars. Vietnam was nearly 'bombed into the Stone Age' without surrendering to the Americans. So was Afghanistan, without giving in to the Soviets or later to NATO. The advent of lasers, miniature computers, smart weapons, and other technical improvements has not obviated the need for line infantry employed much the same way that Roman legionaries were thousands of years ago. However, such empirical evidence has not shaken the belief that technology has fundamentally altered the nature of war and that technology will solve military problems:

> In the West, the cornucopia of novel technologies has meant that we have increasingly sought a technical answer to every security concern. If we devote increasing resources to the expanding range of opportunities, we may find ourselves less and less able to afford to procure the weapons that we need for our security. This is the Technology Trap, and it has as its bait 'the neat solution'. Just like the cheese in a mousetrap, the apparently free benefit may carry a hidden and terminal cost.[27]

It is not a case of abandoning technological improvements to weaponry; it is that the technology should be appropriately understood and used for what it is: an improvement in how something is accomplished, an aid, a refinement.

THE THEORY OF EVERYTHING

Information Warfare, 4GW, MW and EBO are all good ideas, but they are not manifestations of an RMA that has altered the nature of war in some fundamental way. They are, instead, manifestations of the influence of technology. They are nothing new; they only appear new to the extent that an improvement in some aspect of warfare, or the rediscovery of a tool or a method, has given rise to some euphoric prediction of RMA. History is a great teacher in this respect, but Clio must be given time to perform her educative function.

On the question of whether our various military theories have evolved over time, Gould presents a compelling argument for the negative. Western society now firmly believes that the evolutionary process has had a well-defined tendency towards amelioration and refinement since shortly after Darwin introduced his concept of natural selection. But faith is not the same as proof. It *may* be so, but Gould has demonstrated that we have been living with this error of logic for so long that it has become an accepted truth. Starting from first principles, it should be possible to demonstrate the premise. That there has been change is clearly provable and has been proven repeatedly. That this change is always towards amelioration is only an assumption, one which has often been misinterpreted as fact.

Such misinterpretation is a common trap. As the Armour Directing Staff at the Canadian Army Command and Staff College in the 1990s, I would regularly ask students to explain enemy's intent. Invariably they would reply that the enemy stood with combined arms formations forward and with armoured and artillery formations in depth and so on. To this the students would always hear the same response, that what they offered was the enemy disposition. What I wanted was his intent. It is a simple mistake, which is easy to make, and it remains widespread amongst army officers. Widespread belief is comforting but it does not always prove correct. Reading Sun Tzu might have helped:

> The ultimate in disposing one's troops is to be without ascertainable shape. Then the most penetrating spies cannot pry in nor can the wise lay plans against you. It is according to the shapes that I lay plans for victory, but the multitude does not comprehend this. Although everyone can see the outward aspects, none understands the way in which I have created victory.[28]

The fallacy, which Gould exposes, is widespread in military circles. The latest military theories mentioned above are held up as paragons of virtue to save present and future armed forces from the blundering of the past. However, there is not proof that these military theories are a better evolutionary example than those that they wish to depose. There is no evidence that mankind has a fundamentally better understanding of the nature of war today than it did one hundred or even one thousand years ago. In fact, there is evidence that with Western society's increased seduction by dazzling technology, it has lost ground and begun to lose some of the understanding that had previously been gained. Naturally, we do many things better now than before. But the latest theory – pick one – with its claim that what has gone before needs to be swept away as the detritus of history, contains a fatal internal flaw: its wisdom is drawn from the past. This wisdom of the ancients, whether Chinese, Roman, or Greek, continues to resurface as the truth that is too often ignored. The Theory of Everything may, indeed, be close at hand and, when it appears, it may well bring about an RMA. However, until it appears, we will continue to work with old ideas, recombine them into new ideas and continue to look for some better way to describe something that has remained beyond full comprehension thus far.

Notes

1. Robert S. Bolia, "Overreliance on Technology in Warfare: The Yom Kippur War as a Case Study," Parameters (Summer 2004), 46.
2. Ibid.
3. Jon Sumida, "Pitfalls and Prospects: The Misuses and Uses of Military History and Classical Military Theory in the 'Transformation' Era," *Rethinking the Principles of War* (Annapolis Naval Institute Press, 2005), 127.
4. Colin S. Gray "How Has War Changed Since the End of the Cold War?" Parameters, (Spring 2005), 15.
5. Williamson Murray, "Thinking About Innovation", Naval War College Review, (Spring 2001), 121.
6. Bill Keller, "The Fighting Next Time", The New York Times Magazine, March 10, 2002.
7. See Roman Jarymowycz, "The Quest for Operational Maneuver in the Normandy Campaign: Simonds and Montgomery Attempt the Armoured Breakout", PhD diss., (McGill University, 1997).
8. Murray, "Innovation," 119.
9. Ralph Peters, "After the Revolution," Parameters, (Summer 1995). 7.
10. Victor Davis Hanson, "The Western Way of War" Australian Army Journal, (Winter 2004), 159-160.
11. Fritz Sternberg, *The Military and Industrial Revolution of Our Time* (Westport, CT: Praeger, 1959)
12. Paul F. Braim, "An Earlier Revolution in Military Affairs." Parameters (Autumn, 1996), 151-54.
13. Williamson Murray, "Innovation," 127.
14. Keller.
15. Ryan Henry and C. Edward Peartree, "Military Theory and Information Warfare", Parameters, (Autumn 1998), 121.
16. Patrick Lambe, "The Perils of Knowledge-Based Warfare," available from http://www.destinationkm.com/articles/default.asp?ArticleID=1043.
17. Peter Paret, Innovation and Reform, 11.
18. Stephen L. McFarland, "Battles Not Fought: The Creation of an Independent Air Force", Harmon Memorial Lecture No. 40, (Colorado Springs, CO: United States Air Force Academy, 1997), 3.
19. *Operation Desert Storm: Evaluation of the Air Campaign* (Letter Report, 06/12/97, GAO/

NSIAD-97-134), available from http://www.fas.org/man/gao/nsiad97134/letter.htm.
20. Henry and Peartree, 133.
21. Robert Scales, "Human Intel vs. Technology: Cultural Knowledge Important in Iraq" Washington Times (February 3, 2005).
22. *United States Marine Corps Operational Handbook 9 3 (Rev A) Mechanized Combined Arms Task Forces*, (March 1980) paragraph 202 as quoted by B.G. Brown in "Maneuver Warfare Roadmap Part 1", Marine Corps Gazette, (April 1982), 43 (stress in original).
23. A.A. Vandegrift as quoted by William Lind in "Why the German Example", Marine Corps Gazette, (June 1982), 60 (Emphasis added).
24. Brigadier General M.J. Ward, Commander of Canadian Army Land Force Doctrine and Training System commenting in a Trip Report on what he observed during Command and Staff exercise VIRTUAL RAM 24-25 JAN 05. The reference to the British study of Operation Kinetic is the one completed by Lieutenant Colonel J.P. Storr, The Command of British Land Forces in Iraq, March to May 2003. Directorate General of Development and Doctrine, British Army, Spring 2004, Wiltshire UK. In his study LtCol Storr explains that British units have become obsessed with the OPP, forgetting that the whole point is to command units to fight. The commanders and staffs of higher formations in effect were only concerned with themselves, not issuing orders in any timely or meaningful way to subordinates.
25. J.P. Storr, *The Command of British Land Forces in Iraq, March to May 2003* (Directorate General of Development and Doctrine, British Army, Spring 2004, Wiltshire UK).
26. Cyril Falls, *A Hundred Years of War*, (London: Gerald Duckworth & Co., 1961), 13.
27. Timothy Garden, *The Technology Trap*, 6.
28. Sun Tzu, *Art of War*, 100.

" Prediction is difficult, especially about the future.

―――――

Lawrence 'Yogi' Berra

THE FUTURE OF WAR

Introduction

Why has more than two millennia of investigation brought Western society only marginally closer to understanding war's true nature? The investigation thus far has been backward looking. We now reverse our gaze and look forward along the arrow of time to try to pierce the veil of a hidden future. Using three examples, we shall see that only by using the known past as a rudder can we steer a safe course into the future. For even though it may be true that the past is a poor guide to the future, it remains the best tool we have.

Knowing that the past informs the future, the question arises: 'Whither warfare?' Will the prophesies of bloodless war ever be fulfilled? Will one of the current popular military theories – 4GW, Cyberwar, MW or EBO – become the new baseline military theory for our collective understanding of the nature of war? Will mankind destroy itself in some nuclear conflagration as so many science fiction writers have predicted? Will the United Nations finally achieve the vision of a world government, which will obviate the need for war? Will Bernard Brodie's vision of armed forces that work primarily to avert wars instead of to fight them be realized?

The safest method to ponder the future is to consider trends (keeping in mind that trend analysis is notoriously unreliable). Advances in information technology will undoubtedly continue; the spread of the knowledge and materiel required to build weapons of mass destruction (WMD) will continue; democracy, although appearing to be on the rise, which ensures peace in the First World, suggests political turbulence in the Third World where stable democracies are relatively rare. All these factors and more form part of the daily changing situation with which military planners must contend.

Debate regarding the future direction that warfare will (might) take remains continuous. Pundits abound. Why is this debate so animated? For many, what is observed no longer fits inside the well-defined boundaries of conventional warfare. The long-accepted models, used for generations to help in our collective understanding of the nature of war, no longer satisfy. Forcing the empirical evidence to fit an inappropriate model may be worse than having no model at all. Donald R. Baucom offers an insight:

> In *The Structure of Scientific Revolutions*, Thomas S. Kuhn proposed that the progresses of a given science are ruled primarily by its paradigm, an intellectual framework, which comprises the knowledge of that science and the rules governing the conduct of its study. The paradigm forms the scientist's view and dictates the research questions he might ask, thus determining the direction in which the science will develop. Periodically, explanations of observed phenomena within the paradigm become disturbing. The phenomena might still be explicable within the paradigm; but the explanations may be so complex as to lose their validity.[1]

The old 20th century paradigm of war may now be invalid. Conversely, forecasted future ones may likewise be of no use. For the same reasons that the best peacekeepers are highly trained professional soldiers, the best guarantee of peace is the existence of professional standing armed forces – the admonition of US founding father James Madison notwithstanding. The Cold War never became a 'hot' war because the US, and its allies and the USSR, and its allies, each had enormous, professional, and deadly military forces poised to strike twenty-four hours a day, seven days a week. Had either side unilaterally disarmed, there was little assurance that the other side would not have taken advantage of the situation. Consequently, the stalemate forced a stable, if uneasy, peace. Certainly, the hundreds of small wars and conflicts fought almost continuously are violent proof that the model is far from perfect. Nevertheless, it is axiomatic that professional soldiers are the first to abhor war for they, better than any amateurs, academics, or politicians, know the horrors of combat.

This idea is hardly new. Vegetius admonishes those who would desire peace to train for war (*Qui desiderat pacem, bellum praeparat*). But it is an idea that begs to be continually relearned. The underlying premise from Vegetius is that thinking about the future is a necessary part of planning for military success. Although verging on the tautological, there is truth in the argument. What makes Yogi Berra *isms* humorous is that they are invariably based on a reality as might be perceived through the eyes of a child. At a well-known New York restaurant: 'No wonder nobody comes here – It's too crowded.' When asked for directions to his home: '...and when you get to the fork in the road, take it.' Humour aside, of course, it is difficult if not impossible to predict the future. But one of the advantages of any theory is that it offers a predictive function.

Therefore, with no specific military theory as a basis for this predictive function,

THE FUTURE OF WAR

this chapter considers the future of war. No claim is made that all future threats, technologies, or types of conflict will be discussed. For example, low-intensity conflicts, peacekeeping, and peace support operations are given no space. A multitude of future trends and scenarios have been predicted including worldwide terrorism, nuclear holocaust, hyperwar, biological and germ warfare, cyberwar, global anarchy, and the 'Clash of Civilizations'. The list is practically endless. To consider them all would be both impractical and not entirely useful, so the field has been narrowed. Technology and cyberwar will be the primary area of consideration, followed by brief overviews of what the 'Clash of Civilizations' and the future of air power might bring. Using these three subject areas as frames of reference, I will offer some thoughts on the future of war and the implications considered from the perspective of war-centred Military Theory.

Consider the Venn diagram below: Three examples illustrate some, but not all, of the possible permutations available to describe the relationship among various potential futures. In the example, the deliberation is among the Future of Warfare, the Future of Force Employment, and the Future of Conflict. As discussed in 'The Utility of Theory', a certain inexactitude of language haunts our discussions. When considering the future of war, it is easy to confuse or conflate the above three terms and discuss conflict when warfare is meant, or warfare when force employment is meant. It is difficult enough to peer into the darkness of the future, even with the illumination of the past as an aide. The process becomes even more difficult when what we see is not understood. Future force employments, future security environments, the future of war, the future of warfare and the future of conflict are all important issues, and they are all closely related concepts; but they are not interchangeable, and they should not be used as such.

An example may clarify. In 1999 General Charles Krulak, then USMC Commandant, wrote an article entitled 'The Strategic Corporal: Leadership in the Three Block War' in which he described what the future of warfare for the USMC might entail. He postulated a theoretical future where, within a three-block urban space, US Marines might simultaneously fight a conventional war, feed starving children in a humanitarian relief operation and be both peacekeepers and peace enforcers. Krulak was attempting to get the Corps to appreciate that the future was likely to be more complex than the present. He stressed that to meet these challenges the USMC needed to reinforce its historical strength of training leaders to be flexible and to be of strong character, while not disregarding future technological advancements. The Commandant thereby, and almost inadvertently, created the concept of the 'Three Block War', which was soon added to the coda of both American and Canadian doctrine – and, incidentally, misunderstood by most soldiers. The subtleties of the misunderstanding could be discussed for many pages. Simply stated, for Krulak, the three 'blocks' within which a soldier would fight, was a metaphorical unity. The Canadian army, however, took each of the three scenarios and labelled them 'Block 1',

THE FUTURE OF WAR

Conflict

Force Employment

Force Employment

Conflict

War

Conflict

THE FUTURE OF WAR

'Block 2' and 'Block 3' referring to them discreetly. This interpretation demonstrated a fundamental misunderstanding; Krulak was introducing a *concept*, not labelling three discreet 'blocks.' The key point was that this misunderstanding was a typical example of bright and educated people (Canadian army doctrine writers) confusing concepts when considering the future of warfare.

This confusion is nothing new for either the Canadian army or several others. The 'grafting' of doctrinal concepts taken from elsewhere on to native doctrine is fraught with risk. Doctrine *must* stem from culturally inspired precepts. *Auftragstaktik* and *Schwerpunkt* work for the Germans because they are *German* concepts. Remember Professor Bill McAndrew admonition during his lectures to Canadian army officers at the Canadian Army Command and Staff College: 'Cactus will not grow on the tundra – no matter how much you care for them.' Naturally, there are conditions under which ideas, tactics and doctrine can be successfully transplanted across national boundaries as was done even as far back as Maurice of Nassau and Gustavus Adolphus. But such 'grafting' must be done with intentional adaption and comprehensive care to consider cultures (both civil and military), time and space.

Obviously, there are inherent advantages to thinking about what the future may bring. Rather than facing unexpected challenges unprepared, those who consider where current trends might be leading become, at least partially, prepared for what may occur: carrying an extra blanket in the trunk of the car in winter; buying a generator in case there is a prolonged blackout; keeping a financial reserve for unforeseen expenses. These are all logical measures that help people navigate the maze of everyday life. In the military, such preparedness is a necessary sub-set of planning and fits under the rubric of 'war gaming.' The bellicose term disguises a simple truth: this procedure is little more than the game of *what if* writ large. But it is an exceedingly useful enterprise. The Prussian General Staff traditionally had an annual theoretical problem that all staff officers had to solve. The US Navy has had a similar tradition in their Naval War College. These problems would not be purely hypothetical. Inevitably, there would be a real-life benefit: 'How does Prussia best defend against a simultaneous attack from France and Russia?' 'Can the USN simultaneously meet the threat of a challenge from Great Britain in the Atlantic and Japan in the Pacific?' 'How long might it take to split the fleet and redeploy the squadrons to different oceans?'

Several decades ago, these intellectual exercises moved out of the realm of the tactical, the operational or even the strategic and into the realm of the philosophical by asking what a future war might look like. In this arena, the American military led the way. There are obvious reasons for this leadership. As the lone superpower (some would say hyperpower) with worldwide responsibilities, planning for future global stability falls to them. After 11 September 2001, America saw itself as a society under siege and therefore focusing on future warfare became an increasingly important part of American military culture. All these factors have had direct influences upon the American military. Since they do lead the pack, students of Military Theory, as well

THE 'FOG OF PEACE'

POTENTIALLY USEFUL	BLINDING FLASH OF THE OBVIOUS	QUESTIONABLE	AMBIGUOUS AND DANGEROUS	DEFIES DESCRIPTION
Revolution in Military Affairs	Asymmetric Warfare	Network Centric Warfare	Information Dominance	Net War
	Effects Based Operations	Omnipresence	Information Warfare	Cyberwar
	Intelligence Preparation fo the Battlefield			
	Predictive Battlespace			

as practitioners of military arts and sciences, are well-advised to understand what the practical implications are or what they might be and there is a real need to keep their fields of vision as broad as possible. In other words, America's military leadership, and which theories inform it, is important to the whole world.

It is not unusual for great military powers to form or influence a great deal of the Military Theory in their contemporary world. Success fosters emulation. The world's armies copied the Macedonians after Alexander conquered the known world just as later cultures copied Rome, France, Prussia, and Britain respectively. In warfare, the 20th century has been America's century. Quickly recovering from a disastrous civil war in the latter half of the 19th century, a reconstituted United States rose after the First World War, first to a position of recognized, if isolated, strength and then, after the Second World War, to a position of military, political and even social dominance not seen since imperial Rome. Even with the rise of China, this trend has shown little sign of abatement. America's annual military budget continues to be larger than the next dozen or so militaries *combined*. During America's steady climb to world hegemony, many states, both friend and foe, began to adopt US methods, thoughts, and concepts on war. Whether it was something simple like adopting American uniforms and mannerisms or something more profound, like doctrine and structure, the most intellectually insidious of these adoptions has been the American belief in the power of technology to bring to heel the dogs of war. In general, America and its military have had a love affair with technology. Scientific approaches dominate almost all issues irrespective of their nature. This trend has been more than a century in the making, and if any change has been discernible at all, it has been the acceleration of the dependence of the American military, followed by almost all other modern armed forces, upon technological solutions to what are at heart human problems.

The first step in finding potential solutions to future problems is to define them. In this respect, language and conceptualization are the chief tools. I have already commented regarding our looseness of language; but what about the language that has been used regarding trends and the future of warfare? Sadly, it does not inspire confidence. Glenn Buchan at the Rand Corporation referred to the language contained

THE FUTURE OF WAR

in the chart below as the 'fog of peace'. As he pointed out in a 2003 Rand study, for all its devastation, war can bring clarity and a sense of purpose; whereas peace often induces "nonsense that constipates military bureaucracies in peacetime."[2] You cannot avoid the feeling that when such language is tossed about almost indiscriminately, theory is relegated to a back seat behind the newest technological innovation.

The belief in technology as a panacea may be both widely held and a dominant view, but it is not universal. The series of wars and campaigns fought by the US after the Cold War – the two Gulf Wars, Kosovo, Afghanistan, and Iraq – has exposed a deep schism in the country between traditionalists and reformers:

> The former basically take a business-as-usual approach to the future, implicitly arguing that evolutionary improvements to existing weapon systems, military strategy, and operational art offer the best approach to meeting future U.S. security needs. The latter believe that fundamental changes in the world and advances in a number of key technologies both pose challenges to the traditional approach and present opportunities for doing things differently and more effectively in the future. The prescriptions of these "reformers" would generally lead to a different emphasis in the selection and design of weapon systems, changes in organizational structures, and alterations in basic military strategy and tactics.[3]

In other words, the two sides of the argument are that the nature of war is immutable, or the nature of war has fundamentally changed because of technological innovation. Even this separation is misleading, however, for the two camps have internal divisions. There are wheels within wheels. The winners inevitably are those with political connections – until an election changes the gameboard anew.

The devotion to technology, particularly in America, manifests itself as several recurring themes. The most common, and the most disturbing, is the conviction that future technology will lift the 'fog of war', the modern iteration of the Duke of Wellington's comment that the principal challenge of the 'business of war' was in guessing what was on the other side of the hill. The phrase itself, 'fog of war', is one of the most ubiquitous in military literature. Although widely attributed to Clausewitz, he did not use the term in the sense we inevitably see it, save once. There seems to be a near-obsession with lifting this fog; that we may 'see' the situation more clearly. In an era of addiction to instant gratification, where anyone with a computer or even just a smartphone can look up information that used to require a visit to a reference library and several hours of diligent research only amplifies this obsession. This need may seem harmless to the general population, but to military leaders, this desire for instantaneity coupled with a reliance on technology may be dangerous in future wars.

Back in 1995 US Admiral William A. Owens published several essays on the theme 'Dominant Battlefield Awareness' and how it would dissipate the fog of war.

Owens' meaning was that by connecting largely existing sensors and shooters together via appropriate information and command-and-control systems, it should be possible to detect, track, and classify most or all the militarily relevant objects moving on land, sea, air, or in space within a battlespace of some 200 nautical miles square. Like so many attractive notions, his caught on, especially with technocrats and soon took on a life of its own. Essays and articles began to appear, and many US officers and authors began to adopt the idea that battlefield awareness equated to battlefield dominance, which is not true. Recall my analogy of the chessboard above. Both players can see all the pieces; but that is no guarantee of victory for either player. Position, although a key component of the commander's battle problem is but a single piece of the puzzle. A far more important piece is the enemy's intent. Discovering an enemy's intent has nothing whatever to do with machines or technology and everything to do with analysis, intuition, and judgement.

Certainly, the technological advances in optics, surveillance, computers and so on have made hiding in any battlespace far more difficult. But even after decades of evidence to the contrary, many continue to pin their hopes on technology to lift the fog. Whether these folks support 'net-war' or some other form of high-tech wizardry what they fail to understand is that although technology is an important combat multiplier, there is no technological panacea. Satellites, sensor arrays and computer networks may help the commander see parts of the battlespace hitherto kept dark, but uncertainty will forever remain in the military and this uncertainty make life difficult for policy makers.

Understandably, the predominant and overwhelming concern during the Cold War was nuclear holocaust; both sides had intercontinental nuclear weaponry as well as battlefield nuclear missiles, bombs, and artillery. Both sides were serious about using them. Military staffs routinely planned for, and practised, the nightmare scenario of nuclear war. Nuclear planning was a curriculum requirement at all NATO staff colleges. The overriding concern, on both sides of the Iron Curtain, was surviving a surprise initial strike by the other side. Mutual Assured Destruction (MAD) – the guarantee by each side that it would strike back devastatingly – quickly developed from a threat to a national policy, to the preferred method of deterrence.[4] In the West, this deterrence led to reliance upon having the Soviets believe that NATO (primarily the Americans but also the British and the French) would destroy them from a continent away as punishment for an act of unwarranted aggression. The threat of nuclear annihilation became a geo-political lullaby that acted to keep both sides calm.

To make the MAD threat credible, command and control systems had to be survivable. Attack plans had to be 'failsafe.' These needs led to elaborate, fully scripted and inflexible war plans that raised the spectre of First World War mobilization schemes that, once launched, would lead all of Europe inexorably into war. Concurrently, with declining Western birth rates, NATO manpower deficiency (real or perceived) led to an increased reliance upon technology to offset a lack of recruits. All these ingredients

THE FUTURE OF WAR

combined to form a high-octane brew of anxiety and deterrence that kept nuclear planners occupied for over four decades.

The complexities of modern life were mirrored in the science of war. The new science of systems analysis, led by the US Air Force, helped unravel the tangled yarn of nuclear policy. Systems analysis enabled men to devise such a typically 20th century proposition by Thomas Shelling that the worst that an adversary can do to us is not necessarily the best that he can do for himself – a concept that underlay the theory of deterrence. Then, in November 1989, the fall of the Berlin Wall and the speedy collapse of the USSR broke with the past. It removed the threat of a superpower nuclear exchange while replacing the tidy Cold War past with a maelstrom of uncertainty. The technology that supported MAD inadvertently moved nuclear warfare from the realm of detente into that of terrorism and guerrilla war. In the US:

> ...the Army basically got out of the nuclear business. The Navy retained the dominant strategic role, but its aircraft and surface fleet shifted entirely to conventional weapons. The Air Force retained vestiges of its nuclear capability – a few hundred ICBMs and some bombers with a dual nuclear-conventional mission – but nuclear weapons moved from its basic raison d'état operational [to] operational and institutional backwater.[5]

Thus, the threat of nuclear holocaust diminished as a concern for prognosticators of war while it simultaneously became a concern, not as a war to fight, but as a terrorist methodology to pre-empt. The end of the Cold War changed the international security paradigm. The half-century after the Second World War had been subject to a stable, if precarious, security environment. It was ordered. It was tidy. The two major superpowers had their allies and their proxy states, all of which engaged in a constantly evolving *toreador* dance of equilibrium politics. In the continual quest for the upper hand, whoever could accurately predict the future, even in the short-term, could gain a decisive advantage.

Electrons vs Blood

Cyberwar, or Net War, though ill-defined terms that are frequently used interchangeably, usually refer to warfare where human beings no longer fight each other directly. However, at its core the conduct of war has always been about human beings hurling things at each other with or without mechanical assistance. War has been about breaking things and killing people; but will it always be? "The information revolution implies the rise of a mode of warfare in which neither mass nor mobility will decide outcomes; instead, the side that knows more, that can disperse the fog of war yet enshroud an adversary in it, will enjoy decisive advantages."[6] It is a beguiling thought. Some authors believe that warfare may soon leave what has been called 'human space', that which is discernable to the human senses, and enter the realm of weapons that autonomously seek, identify, and destroy targets in all three dimensions.[7] Others believe that we may already be there:

> Future generations may come to regard tactical warfare as properly the business of machines and not appropriate for people at all. Humans may retain control at the highest levels, making decisions about where and when to strike and, most important, the overall objectives of a conflict. But even these will increasingly be informed by automated information systems. Direct human participation in warfare is likely to be rare. Instead, the human role will likely take other forms – strategic direction perhaps, or at the very extreme, perhaps no more than the policy decision whether to enter hostilities or not.[8]

Since the dawn of conflict, human beings have been searching for a weapon that would render them invincible in the face of their enemies. Stones and clubs evolved into muskets and cannon; slings and arrows became jet fighters and ballistic missiles. In 1945, this quest reached an almost oxymoronic level as the United States unleashed upon Japan two atomic bombs, weapons believed by some to have been the epitome of the quest for invincibility.

Is Cyberwar, Net War, or Information Warfare the next step? The *Imams* of information technology would have us believe that humanity has already come to the end of warfare as we have known it, that the future has broken from the past, that kinetic energy warfare has given way to digitization warfare. Is this true? If we view the history of warfare correctly, as a continuum, it is obvious that there have been drastic changes over time. Nations no longer meet each other on the field of honour with banners and heralds. Warfare has become infinitely more complex, incredibly more complicated, and global in scope. Computers and satellites control long-range weaponry possessing near-apocalyptic properties. Whereas battles once could be expected to be fought on a few acres of farmland, as was the case with the Battle of Antietam, where in the Autumn of 1862, almost 23,000 Americans died or were wounded on a mere 30 acres, now entire continents are foreseen as killing fields with millions of lives at stake. However, we must appreciate that this change describes the scope of warfare and not the nature of war.

Unfortunately, some proponents of the new model for war often lose touch with reality. It is easy to overlook the fact that although information is an asset and a weapon in war and always has been, it is not the *medium* of war. To use Clausewitz's phrase, "war does not have its own logic, but it does have its own grammar." Violence is the medium *within* which war lives, not the purpose. The table below, adapted from James J. Schneider's essay "The Theory of the Empty Battlefield" offers an elementary comparison of the changing paradigms and domains of warfare past, present and future.

THE FUTURE OF WAR

REALMS OF WARFARE

PATTERN	DOMAIN	EFFECT	OUTCOME
Attrition	Physical	Annihilation	Surrender
Maneuver	Logistical	Exhaustion	Disintegration
Net War	Cybernetic	Paralysis	Incapacitation

In other words, the old paradigm was attrition, where combatants fought each other physically until the bloodletting quite literally reduced one side to the point of surrender, in some rare instances completely eradicating one side. The new paradigm implies victory without human bloodletting. Mass is no longer a key determining factor.

The claim is that Net War (Cyberwar or Info War) represents the most likely path that warfare will take. Predictions foresee warfare where humans do not necessarily ever meet one another; where the desired effect of combat is the paralysis of an entire society; and where the aim of battle is to remove an opponent's capacity to carry on the fight – either as an opponent or, in extreme cases, as a whole society. As with so many predictions, this one, too, is a tempting prospect. The risk is that none of this talk about future warfare, so dominated by technology, talk that seems to make the past irrelevant, will serve well the needs of policy and doctrine. Rather, it will hamper leaders in their understanding of the need to create national policy, strategy and military doctrine based upon a thorough understanding of Military Theory.

The table above can be deceiving; you must appreciate that it only deals with war conceptually. Notionally, the change in warfare has progressed away from physical destruction and towards a type of armed conflict that achieves its aims without the necessity of killing. Intellectually, the move has gone from physical to philosophical; from Clausewitz to Sun Tzu. In the 21st century, there has been a consistent and continuing search for wars that are bloodless. High-tech solutions are exceedingly attractive in a country famous for technical invention and innovation. Americans famously prefer to spend ammunition instead of lives, but if promises of precision, and 'clean' killing seem too good to be true, it is usually because they are. In any case, even such 'antiseptic' battles are only bloodless for one side. Bloodless battles are a chimera. War is an activity that has been drenched in blood.

No one wants a return to bloody wars of intentional slaughter. The difficulty lies in discerning where a new concept *merges* with an old one (the new idea of fighting in the cybernetic domain *enhances* the fighting in the previous domain of logistics or the physical), and where a new idea *supplants* an old one (the old domain is discarded in favour of the new domain). Consider a recent proponent of this new paradigm, American retired admiral Arthur Cebrowski:

> One revolutionary notion Cebrowski favors is "network-centric warfare," a concept he attributes to Sun Microsystems, which means hooking ships, aircraft, satellites and ground forces together to create a rich, shared picture of the battlefield in motion. This is warfare as a kind of interactive, multiple-player computer game, and it changes the dynamics of fighting in important ways. For one thing, everything happens faster. For another, the war becomes more about the chase than the kill; the most important asset on the battlefield is not a weapon but a sensor.[9]

The admiral is not suggesting, in this limited example, that network-centric warfare should supplant other kinds of warfare. He is merely suggesting that new tools are available to make fighting deadlier for the enemy while making it easier and safer for us. It is certainly the obligation of every profession to develop new and better methods to better serve society, no less for the profession of arms. You could argue that Cebrowski was continuing in the historical trend of taking an older form of warfare and improving it. Were this the case, then the admiral's opinions would be laudable; but it is not the case. Cebrowski is breaking with the historical past and insisting that war has undergone a fundamental change.

The study of Military Theory begins with the study of history, to interpret the present in its correct historical context. In General Douglas MacArthur's words, history sheds light on "those fundamental principles, and their combinations and applications, which, in the past have been productive of success."[10] MacArthur was a well-read classicist and appreciated that the future obviously stemmed directly from the past and must, therefore, be informed by it. But, as Williamson Murray observes, there appear to be some worrying trends. There is altogether too much talk of future warfare rendering all previous understandings of warfare irrelevant, of embracing the future and disregarding the past:

> The [US] military services, with the exception of the Marine Corps, reflect the attitudes of the American people in being profoundly ahistorical. The "revolution in military affairs" has been to some extent advocated by people who are disturbingly ignorant of history. The emphasis within the services has been, more often than not, on technology and platforms, as embodying in themselves the necessary direction of innovation. But even more distressing has been the re-emergence of the mechanistic, engineering, systems-analysis approach to thinking about future war that so characterized Robert Strange McNamara's Pentagon in the 1960s. The catastrophic result of that secretary of defense's approach was the waging of the Vietnam War by an American

military that consistently refused to recognize the human factor in warfare.[11]

In the final analysis, Murray's concern lies at the crux of the problem with most predictions of future warfare. Too much stress is placed on the technological aspect of what the future might bring. War is a most human endeavour. "Victory in war requires more than gadgets. And it is seldom bloodless. [It requires] *Menschenführung*, one of those clumsy-precise German words, that means leading human beings."[12] Machines, no matter whether made of wheels and belts or microchips and silicon, are an intrinsic aspect of all of modern society. This includes war and warfare; but these machines are the servants of the human will that controls them. More than one military theorist, philosopher or strategist has gone to great pains to emphasise this point: war is the collision of opposing wills.

Clash of Civilizations

Few scholars in the post-Cold War world have enjoyed the impact of Harvard's Samuel Huntington. He coined the term 'Clash of Civilizations' in 1993: Western universalism versus radical Islam versus Chinese hegemony. Huntington's immensely influential essay – at least for Western militaries – declared that future conflicts would occur along fault lines where these three principal civilizations (he delineates eight civilizations in total) come into contact. Without the two superpowers acting as a dual brake on pent-up local aggression, old animosities would resurface and be fuelled by competing worldviews. From 1993 until 2001, Huntington's thesis was widely read but less widely accepted. More often, the socialist historian, Francis Fukuyama, and his essay "The End of History", which declared that all societies in the post-Cold War age would tend towards liberal democratic capitalism, held sway. That was until 11 September 2001, when *al Qaeda* terrorists turned airliners filled with innocent passengers into American symbols of commerce, capitalism, and military power. Suddenly, Huntington seemed prophetic. The social, economic, and cultural tectonic plates of Western society had finally collided with those of Islam as President George W. Bush declared a worldwide and unequivocal war upon *al Qaeda,* their supporters, and terrorism everywhere. Many interpreted this as the first battle in Huntington's clash.

To this day, America remains 'at war' with radical Islam. The ongoing insurgencies in the Iraq and Afghanistan continue to fuel speculation on the roles of American military power and how it should be applied. Nevertheless, whatever the next step might be, conventional forces remain the default military option. Even as the US army experiments with new weapons or pursues high-tech solutions, soldiers and marines continue to sweep buildings as they always have; tanks supported by infantry continue to patrol and fight; helicopters and jet fighters continue to attack targets that have been identified on the ground.

STRATEGIA

In some ways, the situation in Iraq impinges upon several futures at the same time. Terrorism, guerrilla wars and the Clash of Civilizations all come together in Mesopotamia. Fringe groups threaten to plunge the world into general war, either at a constant low state of fighting (Northern Ireland model) or they threaten to trigger a global conflagration (Sarajevo 1914 model). Terrorism, cultural conflict, wars of national liberation and common thuggery all seem to have become networked in a way that appears to thwart the strengths of Western warfare while exploiting its inherent weaknesses.

Problematically, the premise of Huntington's argument appears to contain an inherent flaw. It is also difficult to see how his argument might fit the model of philosophy, theory, and strategy. In some ways, his premise is like the flaws demonstrated by Cold War strategies: they were military strategies not founded upon sound military theory. Furthermore, the cultural 'tectonic plates' that risk colliding are neither monolithic nor homogeneous. Whether we accept a Marxist view (war is economically based), a Clausewitzian view (war is politically based), or a Machiavellian view (war is based on the power of the state), none fit the paradigm described by Huntington. Although professor Howard is surely correct in his view that war is a conflict between societies, it is not necessarily a conflict between cultures. We need a better theory, based upon a philosophy, to make the case. The Clash of Civilizations does not satisfy this necessary structure. What this means, therefore, is that the argument is a sociological one and not a military one. Do societies, cultures or other groups come into conflict with each other? Certainly, they do; but this is not the same as going to war.

Here is the difficulty with Professor Huntington's argument: Someone can simultaneously belong to Islam, be a Canadian or American citizen and espouse Western values, beliefs, and morals. Huntington's premise, using the model created in part one of this thesis, does not follow. There is no Islamic Military Theory in the same way that there is American military theory or Russian military theory. Is it possible that countries that have espoused radical Islam will come into conflict with countries in the West? Some, like President Bush, believe that it has already happened. However, will the West, as a civilization, clash with Islam, as a civilization? It is impossible to predict. Not until Military Theory becomes so homogenous that it can encompass entire cultures can we even attempt to answer the question.

Future of Air Power

The history of air power has been the saga of the promises – both implicit and explicit – of casualty and risk-free warfare. It has been the assurance, among other things, of winning wars without ground troops, that air forces could defeat naval forces alone and that air power is the most dominant force in warfare. These declarations have continued unabated almost from the inception of air power. Although Douhet was the first and arguably the most strident proponent of air power winning wars almost

THE FUTURE OF WAR

unassisted, he has not been alone. Alexander de Seversky in his 1943 book, *Victory Through Air Power*, boasted that modern war had no need of ground troops. Even in the recent past, proponents of air power have held that the bombing campaign in Kosovo was a model for future war. An article in *Technology Review* entitled "The Ascent of the Robotic Attack Jet" is a good case in point; the article promised untold wonders for unmanned aerial vehicles, claiming that they were the future of warfare. If such claims were new, they would be astounding, but they were not. They were as old as heavier than air flight. Moreover, although many similar claims have punctuated the history of warfare in the air, few if any have ever been fulfilled.

The future of air power as it relates to the future of warfare is so closely linked with the US and the US Air Force as to be one in the same. Although all US military services are keenly interested in the future, the USAF, as the service most heavily dependent upon technology, clearly leads the pack. In simple terms, the air force vision of the future has them as leading the fight and, in some cases, fighting alone. Part of this attitude stems from their military culture, part of from their world leadership in airborne technology and part from the nature of war in the air. Whatever the reasons, the USAF has led the other services in the US in the worship of technology and their spending bears this out. And this conviction is hardly new. The popular view of air force generals who were fixated on technology and nuclear warfare at the expense of all other options promulgated by Hollywood movies like Dr Strangelove was not necessarily too far off the mark. In 1960 USAF General Frederic Smith wrote an article in Air University Review that explained how airpower, using nuclear weapons, could quickly and easily defeat communist insurgencies.

From a Military Theory perspective, or even only from that of air power theory, much of what is being written regarding the future of war based on air power has a weak foundation. Using our paradigm of the complex matrix, with philosophy as a foundation, it is easy to see that in the case of future wars being based either solely or primarily upon aerial warfare, the case is built not on any philosophical foundations but almost exclusively upon the promise of new or even yet undiscovered technology. In an opinion piece in *Armed Forces Journal* retired US army General Barry McCaffrey, a highly decorated combat veteran, argued passionately that future wars would not necessarily be won by high tech in a low or no risk environment. McCaffrey summarized what many see as a misguided view of future war, correctly arguing such faith was naïve:

> It seems as if an unarticulated emerging defense strategy is trading the modernization of the US Army for satellite broadband communications, ballistic missile defense, UAVs, and fabulously expensive aircraft with precise munitions that some people are betting will be capable of destroying enemy ground combat forces in a no-risk environment ... History tells us to not bet that future combat will look only like the historical small war models

of Kosovo, Bosnia, Somalia or Afghanistan. We know that future US joint warfighting forces will face fierce battles where we must seize objectives or hold ground. Situations will confront us where thousands of innocents might die in Israel, or Japan, or Korea, or Saudi Arabia, or Eastern Europe if we cannot engage in direct battle. The US joint combat force must deal with a future battlefield where enemy forces must be rooted out of cities, jungles, mountains or from fortified air or seaports. The Army-Marine team may fight fanatical enemy forces in mass dismounted formations attacking in the dark, in swirling monsoon rains, or in snowstorms.[13]

Conclusions

Whether the future brings Cyberwar, a Clash of Civilizations, the abandonment of warfare by any but air forces, or just more of what humanity has always known, one thing is certain: humanity can never disengage itself from this most human of social activities. What then are some of the inherent risks involved in divining the future of war? Well, to turn again to the American example, one of the major risks is that jingoism and slogans will overcome common sense, that to please political masters, military officers will warp definitions and accepted misunderstandings of war to satisfy leaders with only a limited appreciation of any military theory. Consider the comment made by retired Marine General van Riper:

> We don't have a leadership that's involved intellectually. They simply want to will their way to [victory]. They don't want to get involved themselves and help think the way through. They turn to those inside the military, and they give slogans, and they ask them to write to these slogans. And whether the slogan is something like 'information superiority' or 'dominant maneuver' or 'effects-based operations,' these things just kind of fall out as assertions of what we want, and then ask people to write to them. There is no content, so consequently they can't write anything meaningful, but they're being asked, or in some cases being paid, to write, and they write, and they write. And it's terrible.[14]

Prognostication, though widely practised and highly useful, is a decidedly inexact practice. Understanding the consequences of guessing the future is an art, not a science, more in need of a crystal ball than a computer. But when serious and educated professionals offer opinions about the future of war, these opinions sometimes take on the patina of facts and arguments, as they did in France with the *Jeune École*. It is a dangerous game. Therefore, for the time being, the only surety about the future is that no one can accurately predict it. Only by looking back to where it has been can a society accurately recognise the guideposts of the future. The past not only informs the

THE FUTURE OF WAR

present; it also educates the present regarding how to prepare for the future. In many ways, the future of war lies in its past:

> Whatever about warfare is changing, it is not, and cannot be, warfare's very nature. If war's nature were to alter, it would become something else. This logical and empirical point is important, because careless reference to the allegedly 'changing nature of war' fuels expectations of dramatic, systemic developments that are certain to be disappointed. The nature of war in the 21st century is the same as it was in the 20th, the 19th, and indeed, in the 5th century BC. In all of its more important, truly defining features, the nature of war is eternal. No matter how profound a military transformation may be, and strategic history records many such, it must work with a subject that it cannot redefine.[15]

But the past does not offer templates or formulae. The past offers only lessons. If the nature of war is immutable, therefore, it must be the Matrix of Military Theory that changes. Thus, Military Theory and not technology must inform military policy. Whether we talk of theory, strategy or policy, technology must remain the handmaid and not *vice versa*. To do otherwise is to force technology to serve a misguided policy, a flawed strategy, or a losing theory.

Likewise, misunderstanding Military Theory may result in a useless military strategy placed at the service of political objectives. Examples include the insistence that aerial bombardments could bring an enemy to the peace table, as the Germans tried against the British in two successive wars. The most disturbing trend overall, is the mistaken belief that technology can solve human problems. Whether that technology is advanced C4I systems, robotics, or the military theories that mistakenly have been built upon them, like 4GW and EBO, is irrelevant. This trend belies a fundamental misunderstanding of both human nature and the nature of war. There is an unspoken implication that technology decreases the vulnerability to error by commanders and staffs. This presumption is not only untested, but also naïve and potentially fatal. Meretricious technology inevitably fails.

American history offers an excellent example with which to close the discussion. The American strategy in Vietnam to attempt to bring about large climactic 'Napoleonic' battles with an enemy that was fighting a guerrilla war using Mao's 'small fish feeding in the large ocean' strategy was flawed. As US colonel Harry Summers explained in his seminal work *On Strategy*, America's failure was in not appreciating the nature of war as well as inappropriately applying their military theories. The Viet Cong appreciated better than the Americans that their war was a no-holds-barred guerrilla war of the people and that they needed to fight it simultaneously on all levels, political, moral, physical, and not, as the Americans were doing, solely on the tactical, physical plane.

STRATEGIA

Although most obvious in espousing the traditional strategy of the 'big battle' that has characterized the Western way of war, the Americans are not alone in this respect. Western society has been slow to appreciate the paradigm shift. History professor and classist Victor Davis Hanson:

> Today we have reached a paradox in war. Such is the lethal character of the Western way in war that non-Westerners seek to avoid open confrontation with such forces. As a result, we in the West may increasingly have to fight as non-Westerners – in jungles, mountains and cities – in order to combat enemies that avoid our strengths in positional warfare. In consequence, we may not always be able to draw on the Hellenic traditions derived from consensual government of superior technology and the discipline of free soldiers fighting in formation ... Our modern Western societies must not become so educated, so wealthy or so moral that we lose our resolution to use arms in order to protect ourselves.[16]

If the West is to continue to dominate humanity in the art and science of war as it has done, the study of Military Theory, not new technology, should be used to create the roadmap for the future. As professor Edward A. Kolodziej so poetically described it, Western society needs to recognize that it cannot continue to believe that there are simple Roman solutions for "ageless and intractable Greek problems."[17]

Notes

1. See Donald R. Baucom, "Modern Warfare: Paradigm Crisis?" Air University Review, (March-April 1984), 2-3.
2. Glenn C. Buchan, *Future Directions in Warfare: Good and Bad Analysis, Dubious Rhetoric*, and the "Fog of Peace" (Santa Monica, CA: Rand Corporation, 2003), 4.
3. Ibid, 2.
4. Inspired by the writings of Sir Timothy Garden, "Air Power: Theory and Practice", available from http://www.tgarden.demon.co.uk.
5. Buchan, *Future*, 10.
6. John J. Arquilla and David F. Ronfeldt, "Cyberwar and Netwar: New Modes, Old Concepts of Conflict", RAND Research Review, (Fall 1995).
7. See Thomas, K. Adams "Future Warfare and the Decline of Human Decisionmaking", Parameters, (Winter 2001-2002), 57-71. See also David Talbot "The Ascent of the Robotic Attack Jet" Technology Review, (March 2005).
8. Adams "Future Warfare."
9. Bill Keller, "The Fighting Next Time", The New York Times Magazine, March 10, 2002.
10. *Annual Report of the (Army) Chief of Staff for the Fiscal Year ending June 30*, 1935, (Washington, DC: US Government Printing Office) as quoted by Harry G. Summers, *On Strategy: The Vietnam War in Context*, (Carlisle Barracks, PA: US Army War College, 1983),

121.

11. Williamson Murray, "Thinking About Innovation", Naval War College Review, (Spring 2001), 127.
12. Henry G. Gole, "Leadership in Literature," Parameters (Autumn 1999), 1.
13. Barry R. McCaffrey, "Challenges to US National Security: Crusader Essential In High-Intensity Combat," Armed Forces Journal (International), (June 2002), 7.
14. Lieutenant General Paul van Riper to Frontline regarding "professional schooling", available from http://www.pbs.org/wgbh/pages/frontline/shows/pentagon/interviews/vanriper.html.
15. Colin S. Gray "How Has War Changed Since the End of the Cold War?" Parameters, (Spring 2005), 17.
16. Victor Davis Hanson, "The Western Way of War" Australian Army Journal, (Winter 2004), 163-64. Clausewitz's admonition not to become 'soft' (IV: 11) continues to ring true: We do not want to hear of generals who conquer without bloodshed. If a bloody slaughter is a horrible sight, then this is grounds for respecting war all the more, not for making the sword which we wear blunter by degrees out of concerns for our humanity, until someone arrives with one that is sharp and lops off the arm from our body.
17. Edward A. Kolodziej, "Lessons of Hiroshima: A Geopolitical Perspective," Swords and Ploughshares, (Spring Summer 1995), 23.

> "The thing that hath been, it is that which shall be; and that which is done is that which shall be done: and there is no new thing under the sun.

Ecclesiastes

NOTHING NEW UNDER THE SUN

Introduction

Our study has afforded us two benefits: an introduction to the interconnected expanse of military philosophers, theorists, and strategists; and a new intellectual paradigm with which to appreciate these individuals and their works relative to each other. We have revisited some well-worn ideas as well as introduced some new ones. Most importantly, some old paradigms have been disassembled and a new model has been offered whereby philosophy, theory, strategy, and policy although closely connected, can now be understood as belonging to either a taxonomy or an ontology. In either case, we can now see that these components are key constituents of the new paradigm of the Complex Matrix of Military Theory.

As students of strategy, we must recognize military philosophers, theorists, and strategists within the realms where they act. Consequently, although a certain military philosopher may be a soldier, it does not automatically make him or her a strategist. Likewise, it does not make him or her a theorist. The paradigm of an interconnected matrix has been introduced and investigated. Using this model, the theories, philosophies, and strategies of war that form the foundation of our collective understanding we have investigated the nature of war. The investigation has not been exhaustive; it has been exploratory, for this study has been foremost an introduction. The investigation has touched upon the nature of war, the value of theory, the importance of culture, the theories that govern warfare on land, sea, and air. We have considered the influences of technology, and also taken a moment to consider the future.

The premise of this study has been that despite more than two thousand years of active consideration, Western society has gained little in its understanding of the nature of war. The preceding chapters demonstrated the validity of the premise using the methodology of an historical survey taken through the lens of a new paradigm. Further, this paradigm was enclosed within a unique framework. The new paradigm addressed the relationship among philosophy, theory, and strategy. The unique framework was the model of a complex matrix. The matrix comprised the interconnectivity of philosophies, theories, and strategies as well as tactics, doctrine, warfare, and culture. When taken together, all these parts comprised the Matrix of Military Theory.

Like all models, the matrix was useful, but only to a point. It was valuable to express the idea that all philosophies, theories, strategies, policies, and tactics, as well as a myriad of other components of Military Theory, like the RNA within a DNA helix, constantly combine and recombine. The matrix helped explain how military thought has not been linear; that there has been no single 'true path.' But like all models, it has had limitations. The concept of a matrix does not fully demonstrate the linkages of time and space, for instance. Nonetheless, the concept of a matrix of military theory remains a highly useful tool to act as framework within which to demonstrate the premise of the study.

The investigation was broken into several parts. The first part began with a discussion of theory in general and Military Theory in particular. It also contained an exposition of Military Theory as a component of social history. The next part comprised an historical survey of Military Theory regarding warfare on land, on the sea and in the air respectively. The last part investigated the current state of Western Military Theory and demonstrated the tendency of Western military culture to look for simplistic or technological explanations of the complex Matrix of Military Theory. It also offered a glimpse of the future and what war might become.

Connections

The intellectual history of war is the quest to uncover the secret of achieving victory. It is tempting to seek a linear progression of thought from 'Plato to NATO' or to attempt to find some Newtonian cause and effect. There is none. The truth is that there is neither a linear progression nor a cause-and-effect relationship in the development of military thought and theory. Instead, the growing collection of thought and theory, the matrix, consists of a complex array of connected ideas through time and space. From ancient times to the present, from Mesopotamia to the American Great Plains, from studying the Spartan phalanx to pondering the effects of nascent nanotechnology, the matrix connects the past, present, and future.

An important component of this matrix is the linkage among military philosophers, theorists, strategists, and tacticians. The model offered in the 'Introduction' demonstrates how these actors relate to each other. These men read the same books – sometimes each other's. Often, they moved in the same social circles. Frequently, they shared a philosophical *Weltanschauung*. Where they did not, they were the disciples of those who had gone before them. Lloyd served under de Saxe and the Marshal's influence is obvious in his writing. Scharnhorst studied both Lloyd and de Saxe in his *Militärische Gesellschaft* and discussed concepts sometimes borrowed from Archduke Charles, sometimes from Bülow, sometimes from Behrenhorst. Many of the same concepts and ideas resurface both in Napoleon's correspondence and in Jomini's *Art de la Guerre*. Clausewitz was Scharnhorst's adjutant. Can there be any doubt that they both read Lloyd? Sometimes influences, including biases and disagreements, were obvious. Napoleon was quite clear in openly paying tribute to seven great commanders: Alexander, Hannibal, Caesar, Gustavus Adolphus, Turenne, Eugene and Frederick the Great. Sometimes influences were more subtle. Alfred Mahan was clearly a student of Jomini – probably due to his father's influence, even though both Jomini and Clausewitz can be discerned in his work. As far as war in the air is concerned, it is almost impossible to find any air power theory, or theorist, not somehow traceable to Douhet. In short, everyone is somehow linked to everyone.

For a long time, little progress was made in the advancement of Military Theory. For both land and sea, leaders repeatedly studied and considered the campaigns of Alexander and Caesar. The primacy of Vegetius endured for a millennium. In the late Middle Ages, Europeans began to expand their intellectual horizons regarding Military Theory. Machiavelli's writings were a new beginning. The period that followed him saw no new military philosophers of note, but it did see literally dozens of theorists and soldier-scholars. Men such as Montecuccoli, de Saxe, Lloyd, Maurice of Nassau, Gustavus and Frederick all contributed to the growing body of treatises on war. Although each borrowed and built upon some former text, little or nothing new emerged. Soldiers read and re-read the work of Machiavelli and Lloyd. Sailors read of the epic struggles of fleets at sea. The reason for this behavior was obvious: success was imitated and in its imitation lessons became both internalized and eternalized.

The late 18th century saw two substantial shifts in war and warfare. The first was theoretical. The next was practical. The early work done by Lloyd led to a duality among European thinkers, each typified by a Prussian. The one school, following Bülow, moved towards a mechanistic, deterministic approach that searched for war's underlying principles. The other, led by Behrenhorst, tended towards the humanistic, romantic approach that believed in the particularism of war, believing that human ability, unique conditions, and luck governed each encounter. Abler men soon displaced Bülow and Behrenhorst. Jomini presently championed the tradition of precise operational analysis, based on logistics and topography while Clausewitz stressed the emphasis of the uncertain and unpredictable, the predominance of will, personality and moral fibre.

The Wars of the French Revolution fomented the second change. These wars were a watershed in military thought and theory. The French brought about a revolution in the conduct of battle by applying the results of decades of military experimentation and theorizing. This was possible due to France's changed social and economic environment as well as the country's need to defend itself. Thus, the French were particularly receptive to military innovation. This innovation spurred surrounding countries to keep up; arms races were perilous contests to lose. Scharnhorst recognized that France's revolutionary armies had wrought a new method of fighting. He wanted desperately to be able to incorporate some of these methods and ideas into the Prussian army. Prussia's once proud force had ossified; it had failed in its purpose of protecting the nation. In the process of doing so, Scharnhorst and his successors created a new military theory that the rest of the world would eventually emulate.

By the time of the First World War, most Western nations had re-invented their armies, but had they understood the underlying military theory that the Prussians had created? Arguably, they had not. They had damaged, if not broken, the delicate bonds that held philosophy, theory, strategy, and doctrine together. Further, monarchs, presidents and generals all failed to comprehend that the 19th century had wrought a fundamental change in the political structure of their world. The latter half of the century had seen an abandonment of the use of war as an instrument of foreign policy for, even when wars were won, the cost to the nation was too high to bear. The advancements in technology had made these costs escalate exponentially. By misreading the signs that this new technology would make short wars less likely, with the turn of the new century the European powers blindly and mechanically marched into the First World War. They did so in the belief that a few short battles or perhaps one rapid campaign would settle the matter, after which they could all return to their arms races. Instead, the continent mired itself in a four-year bloodletting that bore a greater resemblance to a charnel house than to decisive battle.

At sea, a similar course had been charted and by the end of the 19th century navies finally had theories of their own. Mahan and Corbett had both offered cogent views on how best to use fleets. However, the two theories were not consonant. This was not surprising considering that Clausewitz and Jomini were the primary influences upon the two navalists. In France, *la Jeune École* was an object lesson in intellectual narcissism. Realizing that it could never topple the Royal Navy from its position of blue water dominance, the French navy became enamored of a tantalizing mixture of new technology and *guerre de course*. The idea of technology making up for physical shortfalls may have been an interesting intellectual exercise, but the reality fell far short of the promise. Air power still lacked any cogent military theory and was little more than an extension of land power, placing the commander on a higher hill.

The Second World War allowed the thoughts of men like Fuller, Liddell Hart, and Mahan to be put into full practice. Between the wars, Military Theory had made several great strides. Fuller and Liddell Harts' concepts of deep manoeuvre, brain warfare and the indirect approach, brought renewed interest in Military Theory. Likewise, the advent of air power theory by Douhet excited the imagination of flyers who believed that the airplane could be the ultimate military weapon. At sea, battle fleets looked for Mahanian blue water engagements. The addition of aircraft to flotillas expanded air

NOTHING NEW UNDER THE SUN

power and sea power simultaneously. The Second World War became, therefore, the 'Great War of Manoeuvre.'

After the peace of 1945, the Cold War saw a decoupling of Military Theory from its historical roots. Technology grew rapidly in importance and military theorists we displaced by technical experts, economists, and mathematicians. Luckily, the effectiveness of deterrence theory and mutual assured destruction was never fully tested. But over the course of four or five decades, Western society lost its collective memory regarding Military Theory. The faith in technology helped erode an already tenuous appreciation for the nature of war. Peter Paret put it most succinctly. Although referring to the American public, Paret's admonition was a warning to all modern societies. Societies needed to appreciate "

> ...that not all wars are fought to achieve total military victory [and] that more than ever sanctuaries, for allies and neutrals, and numerous other restricting factors are compelling realities, between which statesmen and soldiers must wend their difficult and dangerous path in search for the best possible political results.[1]

Thus, from Sun Tzu to the 21st century, Military Theory has been a growing matrix of thoughts, philosophies, theories, and strategies that have built upon the past, sometimes adding, sometimes subtracting, sometimes replacing concepts as mankind has sought to better understand the nature of war. In this respect, the art and science of war has been like any other art or science. Master taught apprentice, who proceeded to become a master. In this way, traditions have been perpetuated, lessons have been learned and techniques have been developed.

However, despite Military Theory building upon the past, the student of war cannot help but be struck by the fact that the more we investigate the past, the more the same ideas resurface. Little or nothing is new. Sun Tzu wrote more than 2,500 years ago. Yet, modern students of war still turn to his *Art of War* for insights into war's nature. Machiavelli's *Arte della Guerra* also continues to offer valuable insights – despite being almost five centuries old. The writings of Lloyd, de Saxe, Jomini, Douhet, Mahan, Clausewitz, Moltke *et al* all continue to offer important insights. All contain valuable lessons regarding war's conduct, how to achieve victory, how to avoid defeat. Their philosophies, theories and strategies still merit study and are as valid in the 21st century as when they were first postulated.

Summary and Conclusions

First, if language is used carelessly it acts as a barrier to understanding rather than as a clarifier. Military officers are notorious for their sloppy use of language. A good example is the confusion that arises when speakers use 'war' when they mean 'warfare' and vice versa or interchanging the terms 'nature' and 'character', often misstating that the nature of war is evolving when what they mean is that the nature of warfare is evolving. The nature of war is immutable, its character changes with each conflict. Further, like any professional group, they use jargon that does not always aid in communication. Partly this confusion is perpetuated by the absence of a universal, unified Theory of War. Although it is both inevitable and useful to have national and cultural influences within the greater realm of Military Theory, the lack of a universal understanding of the nature of war combined with the infelicities of language continue to act as dampers on the development of a global understanding of war.

Second, although Military Theory is widely discussed, it is poorly understood. To most military officers, Military Theory equates to quoting Clausewitz. Military thought and theory languish as understudied academic disciplines; those aspects not hijacked by sociologists and anthropologists are studied only sparingly. War-centered Military Theory is studied mostly at select schools like the US Army School of Advanced Military Studies in Leavenworth. Even there, only a handful of officers spend a year investigating the finer aspects of Sun Tzu, Clausewitz, Jomini, *et al*. Such schools do not even exist in Canada or Great Britain. No such school exists in Germany whence comes so much of the source material. It is true that officers are more highly educated than ever before; but their educations have tended to be in the realm of education for its own sake. Large numbers follow courses in business administration, public administration, economics, and engineering. The result is that officers are only thinly educated in their own profession and subsequently lack the depth of understanding required to serve their profession well.

Our study focused on those individuals who *most directly* influenced how war was either perceived or fought, beginning with war on land. Some men were influential by their actions: Maurice of Nassau and Napoleon, for instance. In other cases, the influence was hidden to all but the true student of war: General Lloyd was read by practically all European officers before the Napoleonic Wars but is now practically unknown. There were clear links among these men. Although these links were not always linear, they were always present. In some cases, there was progression of thought from one person to the next. In other cases, there was a synthesis of two or more ideas or theories. In some cases, there was naught but regurgitation. Nonetheless, there is no discontinuity from the past to the present. All their philosophies, theories, strategies, and tactics were connected in the matrix.

We then surveyed naval and air power theory to gain an appreciation of both the influence of land theorists upon the development of naval and aerial warfare as well as the influence that naval and aerial theorists have had upon land warfare.

NOTHING NEW UNDER THE SUN

Sometimes these influences have been obvious and immediate; sometimes they have been subtle and indirect. Whatever the case, the linkages exist, and they are strong. Regarding war at sea, we saw that despite millennia spent on the water, no one articulated a real theory of naval warfare until the close of the 19th century. Mahan's theory caught the world's imagination and then Corbett recaptured it. The former, connected epistemologically to Jomini, and the latter to Clausewitz, continue to be in competition today. No subsequent naval theory has yet displaced them. In the end, the opposing theories were not so widely different and could be viewed as different components of the naval warfare spectrum. Naval theory also offered an early example of what can happen when a technological solution promises more than it can deliver as was the case with the *Jeune École* and the high-speed torpedo boat.

Within decades of the first powered flight, Douhet produced a theory of air power that, much like Mahan, captured the world's imagination. His successors continue to return to the original concepts brought forth in 1921. In that respect, theories of air power are catholic, even if they remain controversial. In one respect air power theory has been unlike naval and land-based theory. It has consistently claimed that air power should and could displace the older members of the Military Theory triumvirate. In fact, during the Cold War, air power theory did displace more classical military theory. But Cold War theories eventually proved themselves to be intellectually bankrupt and the end of the Cold War saw a return to more classical models of military theory based upon philosophies and strategies that were predicated by historical realities rather than mathematical assumptions.

Next, we turned our gaze to the current state of Western military thought and theory. The role of technology, the concept of RMA, and a brief discussion of whether the realm of Military Theory is evolving to offer humankind a better understanding of the nature of war were considered. The purpose was to demonstrate that often the tool has been confused with its use. This confusion has been especially widespread of late. The declaration that technology has altered the nature of war was demonstrated to be questionable at best and patently false at worst. We then established that a new and revised military theory does not change the nature of war any more than having a new power drill changes the nature of holes in wood. The recent intellectually bankrupt doctrine of EBO, where military forces must not restrict themselves to 'kinetic' action on the physical plane was considered as an example of the facile desire to continuously demonstrate the changing nature of war. The latest American fanciful declaration that civilization now finds itself entering the 'Fourth Generation' of warfare was also considered. Both were isolated, if indicative, examples of unsubstantiated claims that are ample proof of Ecclesiastes' declaration that there is nothing new under the sun as well as the old saying that there is a 'certain gullible type of person' born every minute.

Finally, we turned to the future. The number of authors who have opined on the nature of future war was too large to do justice to their collective body of work. Three of the most popular security scenarios were considered: Cyberwar and technology, Huntington's 'Clash of Civilizations' and the future of air power. These three potential futures were not intended to speak for all possibilities. Rather, they were indicators. Clearly, it is impossible to draw conclusions from events that have not occurred. However, it was obvious that most futurists continue to be heavily influenced by technological change. Too often, these individuals see the future only in the light of what changes new machinery might bring. Little consideration is given to the human aspects of war. This lack of consideration is both a grave error and a recurring theme.

In many ways, this investigation has been a warning for all those who splash in the intellectual pool of Military Theory, a reminder of the danger of falling victim to intellectual narcissism. In other words, we must endeavour not to fall so in love with the splendor of our ideas that we fail to measure them against the harshness of reality. The torpedo boat as a replacement for battleships, strategic bombing to replace ground troops, MW as a panacea, and EBO as a promise of more efficient warfare, were all elegant and enticing ideas. However, war remains a harsh and a deadly environment that brooks no errors. It is a cruel and unforgiving situation and elegance does not trump effectiveness. History provides plenty of examples of philosophers, theorists and strategists who lost sight of this simple truth.

This investigation has demonstrated that the multifaceted field of Military Theory still contains untapped study material. Further, there are no short cuts. Military Theory is far more than reading Sun Tzu or quoting Clausewitz or Liddell Hart. The philosophers, theorists and strategists who concern themselves with war need to have their work read, studied, and understood. The study of military thought and theory is enormously more significant than currently accepted. It must be given more weight. Only with more study can society properly evaluate any of the latest promises offered by technological, organizational, or doctrinal change. Leaders of modern societies – particularly liberal Western democracies – must be educated in Military Theory for the potential consequences of ill-informed decisions are too catastrophic to contemplate. This need is not new. Peter Paret, in his 1966 lecture, *Innovation and Reform in Warfare*, declared that there were three enduring obstacles to society's understanding of the nature of war: "an insufficiently educated public; a failure among too many political and military leaders fully to recognize the political nature of war; and the friction between violence and control that is a permanent characteristic of all armed conflict." Professor Paret's conclusions are decades old, and the importance of Military Theory remains marginalized. Ironically, this neglect is due in part to the great popularity of Clausewitz. Military officers and academics speak of him with such reverence that all other military theories have become secondary. He has become 'an author for all seasons.' The study of *On War* has become practically universal among Western armed forces. In many cases, Clausewitz's interpretations of conflict are no longer spoken of as theory; his thoughts have become revealed truth. Paradoxically, we are left to wonder whether Clausewitz is understood. The primacy of this single text has created a grave gap in the Western profession of arms. Professional knowledge that is so narrowly focused is a dangerous situation for both military and political leaders.

A key consequence of not appreciating that Military Theory encompasses much more than Clausewitz is that much of the study in this area has somehow migrated out of the military realm and into the disciplines of the 'soft' sciences. Sociologists, anthropologists, economists, and psychologists all have added greatly to the study of war but, understandably, each of these disciplines, has tended to see the study of war through the lens of its own needs. The great lesson of the Cold War has been that military commanders must continue to study war from the perspective of war centred military theories lest they end up with theories, which are *not* war centered.

Another reason for the misappropriation of military theories by social scientists has been that too military professionals have considered war only as an exception to the normal course of social intercourse, as an aberration that ultimately operates according to its own discrete laws. There is an adage among soldiers that a nation can support either its own army or someone else's. In other words, countries that allow their military strength to atrophy risk invasion and occupation by foreign powers. A similar caution holds for Military Theory. Any nation that does not study military thought and theory based upon national beliefs, culture and history risks being saddled with theories, strategies and tactics that are disconnected to their national realities. In this respect, countries like Denmark and Canada have much experience, having usually simply used the doctrine of their most powerful ally or nearest neighbour. It is difficult to create independent thought regarding war when you live so deeply in the shadow of a military leviathan. In Canada's case, although there have been many advantages to sharing and enjoying the benefits of whatever intellectual renaissance has occurred in the US, it is also well to recall that the American military has not always been particularly adept at learning from historical experience. For this reason alone, it is incumbent upon Canadian officers to study war and to draw their own, independent conclusions.

Perhaps obvious, but just as important, is the conclusion that there is not yet any single theory to explain all the aspects of armed and organized human conflict: some theories are politically based; some are socially based; some are economically based. Thus far, the search for a universal theory has dealt with scores of useful conceptualizations. Yet, no single theory has incorporated all observations into a universal, unified military theory. The need for such a theory is well established. Such a theory would shine light upon the complexity of the matrix's structure. The revelation of interconnections might also help to connect war with the manifold influences that act upon it, influences such as social structure, religion, economics, politics, and geography. Lastly, theorizing about war allows for the creation of a predictive function. Thus, events, triggers, and relationships could be studied to help predict and exploit new trends before they become uncontrolled or destructive. Moreover, since there is no definitive or universal military theory, it is only logical to conclude that there is no way of knowing whether the nature of war has changed or not. It is impossible to know whether the nature of war is even mutable. It is likely that the various interpretations of war are little more than different manifestations of an increasing understanding of its nature. Just the same, even if no unified theory is ever found, there is value in the quest. The search for the holy grail of a unified theory contains the advantage of keeping humanity aware of the horrendous consequences of not understanding war.

Still another reason for the lack of a universal theory is that soldiering is a practical profession. As discussed in the 'Introduction', there is a natural disdain for theoretical knowledge – especially when this theoretical knowledge contravenes well-practised routines and procedures. The famous economist John Maynard Keynes reputedly said that practical men who believed themselves exempt from any intellectual influences, were inevitably the slaves of some defunct economist. Keynes was talking about bankers and businessmen; but military leaders fall into the same trap. Like all professions, the military cannot allow dead minds to keep obsolete ideas alive.

The problem is with the professional military schools and the military staff colleges. It is here that future commanders are supposed to be educated and intellectually nurtured. But Williamson Murray, describing US colleges, demonstrates a widespread tendency:

> The basic problem is that military organizations can rarely replicate in times of peace the actual conditions of war. It becomes increasingly easy, as the complexities, ambiguities, and frictions of combat recede into the past, for militaries to develop concepts, doctrines, and practices that meet the standards of peacetime efficiency rather than those of wartime effectiveness.

There is no other profession in the world whose peacetime efforts represent only a pale shadow of the harsh realities in which its men and women must carry out their true functions – not least that their opponents are trying to kill them. That is why the profession of arms is the most demanding calling not only physically but intellectually. It is also why professional military education has been so profoundly important to armed services in preparing for and waging war. Here lies perhaps the greatest weakness in the current culture of the American military.

With perhaps a single exception, the colleges of professional military education, charged with educating the officer corps for the complexities and ambiguities of the future, are not especially distinguished. In 2000, a very senior officer told an assemblage at a war college that he hoped its students were getting to know their families and playing plenty of softball and golf, as he had himself when he attended that same institution. At least some of the better students were outraged. It is well to remember, as a contrast, that in the interwar period individuals who were to rise to the highest levels in the coming war had been on the faculties of the war colleges.[2]

If the problem is as widespread as Murray proposes, the profession of arms is in trouble.

So, what has more than two millennia of theorizing about war brought to humankind? Is humanity any closer today to understanding the true nature of war than was Sun Tzu or Machiavelli, Clausewitz? A reader could easily be convinced after reading Sun Tzu's *Art of War* that all the secrets of war's nature had been revealed; that its nature, meaning and conduct could be understood. Arguably, the reader would be correct. Likewise, another reader could come to similar conclusions after studying Machiavelli's *il Principe*, Clausewitz's *On War,* du Picq's *Études* or Fuller's *Generalship*. Again, all *might* be correct.

Herein lies the crux of the problem that has taken up the bulk of this investigation. Like the six blind men of Indostan from John Godfrey Saxe's poem, all the readers above may have indeed discovered portions of war's true nature. As certain as we may be of our modern understanding of war and of our ability to harness technology to solve immediate tactical or mechanical problems, the truth is that we have no better grasp of the nature of war than did Sun Tzu several millennia ago. But there has been merit in the quest. Our millennia long collective search has been an intellectual pursuit. Perhaps for this reason we have faltered. Too often, during the study of Military Theory, the process and the product have been confused one with the other. Too often, the mechanics and the mechanisms of warfare have been confused with the qualities that make up war's underlying nature.

Over the centuries, there has been a tendency for newer generations of readers and practitioners to *replace* prior theories with newer ones rather than to *add* the newer theories to those already extant. Addition builds linkages and adds to the complexity of the matrix. Replacement rejects the building of the matrix. When Clausewitz asserts that war must be understood simultaneously on the moral and the physical planes, the astute student appreciates that this does not contradict Sun Tzu telling his reader that war comprises a dichotomy of weak and strong, of ordinary and extraordinary. Clausewitz's writing is simply an interpretive expansion of what Sun Tzu has already said; the former is theory; the latter is philosophy. This is only one of many examples among the treatises discussed, but it is instructive since it informs the collective perception of the Matrix of Western Military Theory.

The need to build the matrix does not mean, however, that everything written by everyone about war is correct, for this cannot be. It means that, for the most part, all the known interpretations of the nature of war can be accepted as having some validity within the context of a greater whole, within some holistic, unified theory yet unwritten. There are many other examples of such phenomena: the gospels of the Christian church do not always agree; the manifold interpretations of the concept of representative democracy are not always in harmony; the theories of the nature of light being both waves and particles disagree and yet happily co-exist. The key to unlocking the secrets of the true nature of war lies in realizing that all these theories, in the language of religion, are ecumenical rather than catholic. Johns Hopkins professor Eliot A. Cohen notes that history gives no comfort to the many able, subtle, dedicated minds that crave finality and certitude; but it does offer training for those who can tolerate uncertainty and operate within it.[3]

Seen in this light, there is hope that the many insights provided by the philosophers, theorists and strategists might form some unified, cooperative whole. The analogy of a coral reef is useful. When viewed microscopically, each disparate theory, philosophy or interpretive text on war is like an individual, tiny coral. It exists discreetly, independently of the others. However, when viewed macroscopically all these theories form a collective, a colony, a web, a matrix, which lives as a single community. So, what has more than two millennia of the study of war brought us? It has brought us these hundreds of corals in the guise of ideas that now must be viewed as a single coalescent and unified theory. This is no easy task. The fact that some have declared that individual coral represents the entire reef – that Clausewitz, or Jomini, or Sun Tzu has revealed the whole and true nature of war – while new and sometimes contradictory ideas emerge, indicates that humanity may not be as close to uncovering the hidden nature of war as it might believe. In this respect, the answer to the question of whether humankind is any further ahead after more than two millennia of study must be no. But as Professor Cohen implies, the study remains not only instructive but necessary.

Notes

1. Peter Paret, *Innovation and Reform in Warfare*, The Harmon Memorial Lectures in Military History, No. 8 (Colorado Springs, CO: United States Air Force Academy, 1966), 14.
2. Williamson Murray, "Thinking About Innovation", Naval War College Review, (Spring 2001), 121.
3. Eliot A. Cohen quoting Jacques Barzun, "History and the Hyperpower," Foreign Affairs, (July-August 2004), 62.

" You see things; and you say 'Why?' But I dream things that never were; and I say 'Why not?'

George Bernard Shaw

Coda

Welcome to the 'Coda', my explanation of the investigation including criticisms, controversies, and explanations. Although it may sound like it at times, this is not an *apologia*. It is a clarification. As with the purpose of the investigation, which was laid out in the 'Preface' (What is presented here is not enough and I strongly encourage any serious student to seek out the original works of all the military philosophers, theorists and strategists covered in this volume.) below, I will offer some deeper explanations and insights, which would have cluttered the investigation had I included them there.

Over the decade that it took to compile this work, I have faced many criticisms, many valid. Most often, I have faced critique based on misunderstandings of what the investigation entailed, in much the same way that Liddell Hart was fond of misquoting the contents of *On War*.

Strategia was developed from my 2006 PhD dissertation. I originally cobbled together three of the chapters as a *précis* and offered it to students in my Norwich University graduate classes in Military Thought and Theory. I had found that the students, even those who had read military history, struggled in understanding the many disparate views on military theory and so I decided to offer them a 'primer' of sorts as an aid in understanding, to whet the appetite, introduce, and guide those seeking a better understanding of the subject. The primer was never intended to replace any of the original works or to become a *Reader's Digest* of military thought and theory. I maintained then, as I do now, that the original works **must** be read.

At one point, I included the term 'intellectual romp' to describe the investigation. Some early reviewers agreed that the 'romp' was entertaining if a bit uneven. The criticism came from the fact that it was not a *complete* history of military thought and theory. Why had some [favourites] been left out? My explanation, as pointed out in the 'Introduction' was that it is not meant to be a reference text. I have no intention of reproducing Azar Gat's, Sir Michael Howard's, or John Keegan's work. Choices had to be made and those choices are personal; but I was pleased to hear that it was entertaining.

In sharing large portions of the text with educators at American training institutions, some concern was raised that my The Complex Web of Military Theory was yet another 'proprietary' model and that such models are explored at the expense of covering primary texts. One criticism was based on the belief that commentaries do not improve on the originals. I disagree. Again, there is a failure to understand what the purpose of the model and the commentaries are. Anyone who has attempted to study physics without some sort of analogous model(s) will attest that it makes for hard sledding. Even Albert Einstein used analogies like an elevator in space or a train approaching the speed of light. The point is that models and analogies do not *replace* any of the original work. The model places them in context and the commentaries assist the novice reader. Annotated texts of original works have long been strong teaching tools and I stand by my methodology.

Several interesting concerns were raised regarding the nature of war. One was that I was not explicit enough in distinguishing between the nature of war and the character of war. It is a fair observation and I feel compelled to explain. In the 'Summary and Conclusions' of Chapter 1, I am quite clear in delineating the difference, raising the metaphor of Plato's Cave: "Rather, they are tools to aid in our understanding of Plato's shadows on the cave wall, of coming to grips with the fact that the nature of war seems immutable while at the same time the character of war changes continuously". The question is whether this is enough of an explanation. Do I need to strengthen the point? I chose not to for one specific reason: The argument is Clausewitzian. Clausewitz states that war's nature is immutable and that every war is unique. I agree with him but that is immaterial. Overemphasizing Clausewitz diminishes the point of the book, which is to cover the entire waterfront of military thought and theory and not to shill for any individual theory. The distinction between the nature of war and the character of war is arguably the least understood aspect of military theory and a *huge* topic. I recently listened to an ex-Commandant of the USMC explain it and he had it exactly wrong. But as a retired four star no one dared tell him so. Luckily, he is not influential – merely a national security advisor to the US Secretary of Defense. Could more be written? Perhaps.

Another criticism, is that I am wrong in my assertion that we do not know the true nature of war. There is a strong belief among many that we already know what the nature of war is. I disagree. The analogy here that I use is like trying to understand the nature of the universe. When Sir Isaac Newton established his Law of Gravitation, many in the scientific world felt that we had finally unlocked the secrets to the nature of the universe, and just as Aristotle had been 'established truth' for centuries, so was Newton's theory. But then scientists like Albert Einstein and Max Planck shook those foundations and scientists like Stephen Hawking and Edwin Hubble gave us entirely new perspectives on the universe's nature

As I write, British scientists have announced that their work at the Joint European Torus (JET) tokamak machine in the English village of Culham, near Oxford has made a small leap toward sustainable controlled nuclear fusion, a process which holds the promise of almost limitless clean energy. Why do I mention it? If we had stopped trying to understand the nature of the universe with Aristotle, or Newton, or Einstein or, Hubble, or Hawking we would still be using wood and coal to heat our homes.

So, do we know the nature of war? Perhaps, but I am not convinced, and it is for this reason that I speak of not knowing war's 'True Nature'. To be clear, I am not saying that war's nature is unknowable. This accusation is a misreading of my argument. I am saying that we have not yet fully accomplished that goal and contrary to some criticisms, I have not established a strawman to burn down. I use the term 'unknown' because the debate rages on in the same way that the nature of light is 'unknown' as is the nature of gravity, this although Newton gave us startling insights. We have many such valuable insights but the definitive description of the hidden nature of war remains open, and 'unknown'. That is not the same as saying that it is unknowable.

The last criticism, a valid one, is that I have not given the reader a good enough diagram of my so-called Complex Matrix. I agree. I have not and so I will describe here why not.

At the conclusion of 'The True Nature of War, I offer the reader a simplified model, model 'C'.

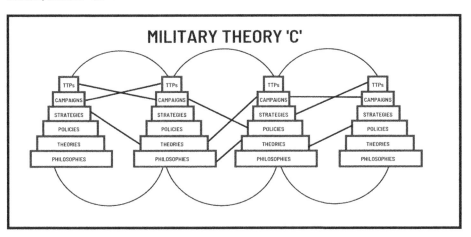

STRATEGIA

It is a very simplified attempt to explain how all of the many facets of military thought and theory connect with each other in an ontology of sorts. There are many, many other ways to look at this issue. Consider the humble but ingenious Rubik's Cube. If we were to make each colored cube a concept, then it might suffice as an intellectual model of the matrix, especially since the cube is really a 'cube of cubes', which we can manipulate thereby changing the relationships among our many concepts, but even that model leaves much out. Perhaps a better model would be a vast neural network which somehow tries to mimic the type of network within our brains.

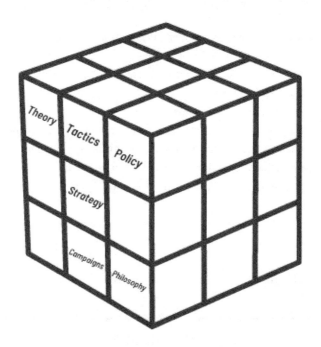

Better, but still lacking something; time and space are left unaddressed. And so, we are left with lots of good but imperfect models and perhaps that is a good place to stop.

"They held a view, still found among some officers, that abstract theories are not useful in the real world of war. They did not realize that much of the operational art is based on abstract concepts, e.g., synchronization, integration, manoeuvre, centres of gravity, etc.

Professor Allan English

Bibliography and Suggested Reading

Books

Alger, John I. *The Quest for Victory, The History of the Principles of War*, Westport, CT: Greenwood Press, 1982.

Beaufre, Andre. *An introduction to strategy with particular reference to problems of defense, politics, economics, and diplomacy in the nuclear age*, R.H. Barry, trans., London: Faber, 1965.

.........*Strategy of Action*, R.H. Barry, trans., London: Faber, 1967.

Bouthoul, Gaston. *La Guerre*, 4th ed., Paris: Presses universitaires de France, 1969.

Brodie, Bernard. *Strategy in the Nuclear Age*, Princeton, NJ: Princeton University Press, 1971.

.........*War & Politics*, New York: Macmillan, 1973.

Chaliand, Gérard. *Anthologie Mondiale de la Guerre Stratégique des Origines au Nucléaire*, Paris: Robert Laffont, 1990.

Chandler, David. *Atlas of Military Strategy, 1618-1878*, Don Mills, ON: Collier Macmillan, 1980.

.... *Napoleon*, London: Weidenfeld and Nicolson, 1973.

Citino, Robert M. *Blitzkrieg to Desert Storm: The Evolution of Operational Warfare*. Lawrence, KS: University of Kansas Press, 2002.

.... *The Path to Blitzkrieg: Doctrine and Training in the German Army, 1920-1939*, Boulder, CO: Lynne Rienner, 1999.

von Clausewitz, Carl. *Principles of War*, Hans W. Gatske, translator, Harrisburg, PA: Military Service Publishing Company, 1942.

.........*On War*, Michael Howard and Peter Paret, editors and translators, Princeton: Princeton University Press, 1984.

STRATEGIA

Colomb, P.H. *Essays on Naval Defence*, London: W. H. Allen and Co., 1893.
Corbett, Julian S. *Some Principles of Maritime Strategy*, London: Longman, 1911.
Corum, James. *The Roots of Blitzkrieg*, Lawrence, KS: University of Kansas Press, 1992.
van Creveld, Martin. *Supplying War*, New York: Cambridge University Press, 1977.
.... *Fighting Power: German and U.S. Army Fighting Performance, 1939-1945*, Westport, CT: Praeger, 1982.
.... *Command in War*, London: Harvard University Press, 1985.
.... *Technology and War*, New York: Free Press, 1989.
.... *The Art of War: War and Military Thought*, London: Cassell, 2000.
.... *The Transformation of War*, Toronto, ON: Collier Macmillan, 1991.
Douhet, Giulio. *The Command of the Air*, Dino Ferrari, translator, U.S. Air Force ed., Washington: DC, 1983.
Dyer, Gwynne. *War: The New Edition*, Toronto: Random House, 2004.
Jomini, Henri. *The Art of War*, 1836 ed., G. H. Mendell and W. P. Craigbill, translators, West Point, NY: U.S. Military Academy, 1962.
.... *Summary of Art of War. 1838*, J.D. Hittle editor, Harrisburg, PA: Military Service Publishing, 1947.
Frederick the Great. *Instructions for his Generals*. Thomas R Phillips editor, Harrisburg, PA: Military Service Publishing, 1944.
Liddell Hart, B.H. *Strategy*. Second Edition, New York: Praeger, 1968.
.... *The Strategy of Indirect Approach*. London: Faber and Faber 1941.
Lind, William. *Maneuver Warfare Handbook*, Boulder: Westview Press, 1985.
Lloyd, Henry. *The History of the Late War in Germany between the King of Prussia and the Empress of Germany and her Allies, Part I Vols. 1 and 2.*, London, S. Hooper, 1781.
.... *The History of the Late War in Germany between the King of Prussia and the Empress of Germany and her Allies, Part II.*, London, T. and J. Egerton, 1790.
.... *The History of the Late War in Germany between the King of Prussia and the Empress of Germany and her Allies, Part I Vol. 1.*, London, S. Hooper, 1781.
Machiavelli, Niccolò. *Il Principe* (1516), Milano: RCS Rizzoli Libri, 1986.
.... *The Discourses* (1519), Leslie J. Walker translator, London: Routledge & Paul, 1950.
Mahan, Alfred Thayer. *The Influence of Sea Power upon History 1660-1783*, Boston, MA: Little, Brown and Co., 1918.
.... *The Influence of Sea Power upon the French Revolution and Empire, 1793-1812, 2Vols.*, London: S. Low, Marston, 1893.
Mao, Zedong. *On Guerrilla Warfare* (1937), S. B. Griffith translator, Fort Bragg, NC: Frederick A. Praeger, 1961.

BIBLIOGRAPHY

Musashi, Miyamoto. *The Book of Five Rings*, Thomas Cleary editor and translator, Boston: Shambhala, 1993.

du Picq, Charles Ardant. *Etudes sur le Combat Antique et Combat Moderne*, Paris: Champ Libre, 1978.

Rommel, Erwin. *Infanterie Greift an, Erlebnisse un Erfahrungen*, Potsdam: Ludwig Voggenreiter, 1937.

de Saxe, Hermann Maurice, comte. *Reveries on the Art of War*, Thomas R. Phillips, translator and editor, *Roots of Strategy: A Collection of Military Classiscs* edition, Harrisburg: Military Service Publishing, 1955. Originally published as *Mes rêveries; ou Mémoires sur l'art de la guerre*, 1756.

von Seeckt, Hans. *Thoughts of a Soldier*, Gilbert Waterhouse translator, London: E. Benn Ltd, 1930.

Thucydides. *History of the Peloponnesian War*, Rex Warner translator, London: Penguin Books, 1972.

Sun Tzu. *The Art of War*, Samuel B. Griffith, translator, Oxford: Clarendon Press, 1963.

Flavius Vegetius Renatus. *The Military Institutions of the Romans*, Thomas R. Phillips, editor and translator, *Roots of Strategy: A Collection of Military Classiscs* edition, Harrisburg: Military Service Publishing, 1955. Originally published as *De Rei Militari, 390 A.D.*

Downing, Brian M. *The Military Revolution and Political Change*, Princeton: Princeton University Press, 1992.

Dupuy, Trevor. *A Genius for War: The German Army and General Staff, 1807-1945*, Englewood Cliffs, NJ: Prentice-Hall, 1977.

.... *Understanding War: History and a Theory of Combat*, New York: Paragon House, 1987.

Falls, Cyril. *The Art of War from the Age of Napoleon to the Present Day*, London: Oxford University Press, 1961.

.... *The Place of War in History*, An inaugural lecture delivered before the University of Oxford, 22 November 1946, Oxford: Clarendon Press, 1947.

.... *A Hundred Years of War*, London: Gerald Duckworth & Co., 1953.

Fischer, Fritz. *From Kaiserreich to Third Reich, Elements of Continuity in German History, 1871-1945*, Roger Fletcher, translator, Winchester, MA: Allen & Unwin, 1986.

Freedman, Lawrence. *The Evolution of Nuclear Strategy*, 3rd ed., New York: Palgrave Macmillan, 2003.

Fuller, J.F.C. *Generalship-Its Diseases and Their Cure: A Study of the Personal Factor in Command*, reprint of 1936 ed., Fort Leavenworth, KS: U.S. Army Command and General Staff College, 1987.

.... *The Conduct of War, 1789-1961*, New York: Da Capo Press, 1992.

Garden, Timothy. *The Technology Trap: Science and the Military*, London: Brassey's

Defence Publishers, 1989.
Gat, Azar. *A History of Military Thought From the Enlightenment to the Cold War*, Oxford: Oxford University Press, 2001.
de Gaulle, Charles. *The Army of the Future*, London: Hutchinson 1940.
Gleick, James. *Chaos: Making a New Science*, New York: Penguin Publishing, 1987.
Goodspeed, Michael. *When Reason Fails. Portraits of Armies at War: America, Britain, Israel, and the Future*, Westport, CT: Praeger, 2002.
Gould, Stephen. *Full House: The Spread of Excellence from Plato to Darwin*, New York: Harmony Books, 1996.
Gordon, Andrew. *The Rules of the Game: Jutland and British Naval Command*, Annapolis, MD: Naval Institute Press, 1996.
Görlitz, Walter. *History of the German General Staff 1657-1945*, Brian Battershaw, translator, New York: Praeger, 1953.
Groener, Wilhelm. *Das Testament des Grafen Schlieffen*, Berlin: E.S. Mittler & Sohn, 1927.
Guderian, Heinz. *Achtung-Panzer: The Development of Tank Warfare*, Christopher Duffy, translator, London: Cassell, 1992.
Gudmundsson, B.I. *Stormtroop Tactics: Innovation in the German Army, 1914-1918*, New York: Praeger, 1989.
Hackett, John. *The Profession of Arms*, London: London Times, 1963.
Hammond, Grant. *The Mind of War: John Boyd and American Security*, Washington, DC: Smithsonian Press, 2001.
Handel Michael. *Masters of War: Classical Strategic Thought*, New York: Frank Cass, 2001.
Hawking, Stephen. *A Brief History of Time: From the Big Bang to Black Holes*, New York: Bantam Books, 1988.
Haythornwaite, Philip. *Invincible Generals: Gustavus Adolphus, Marlborough, Frederick the Great, George Washington, Wellington*, Bloomington, IN: Indiana University Press, 1991.
Heiber, Helmut and David M. Glantz. *Hitler and His Generals: Military Conferences 1942-1945*, New York: Enigma Books, 2003.
Higham, Robin. *Air Power: A Concise History*, New York: St Martin's Press, 1972.
.... *The Military Intellectuals in Britain: 1918-1939*, New Brunswick, NJ: Rutgers University Press, 1966.
Holbraad, Carsten. *The Concert of Europe: A Study in German and British International Theory 1815-1914*, London: Longman, 1970.
Hooker, R.D., editor, *The Maneuver Warfare Anthology*, San Francisco, CA: Presidio Press, 1993.
Howard, Michael. *The Causes of War and Other Essays*, Cambridge, MA: Harvard

BIBLIOGRAPHY

University Press, 1983.

.... *The Franco-Prussian War: The German Invasion of France, 1870-1871*, London: Routledge, 1988.

.........*War in European History*, London: Oxford University Press, 1976.

.... editor. *The Theory and Practice of War*, Bloomington, IN: Indiana University Press 1975.

Huntington, Samuel. *Clash of Civilizations and the Remaking of World Order*, New York: Simon and Schuster, 1996.

.... *The Soldier and the State: The Theory and Politics of Civil-Military Relations*, Cambridge, MA: Harvard University Press, 1957.

Hurley, Alfred F. and Robert C. Ehrhart. *Air Power and Warfare. The Proceedings of the 8th Military History Symposium, United States Air Force Academy, 18-20 October 1978*, Colorado Springs, CO: United States Air Force Academy, 1979.

Iggers, Georg. *The German Conception of History: The National Tradition of Historical Thought from Herder to the Present*, Middleton, CT: Wesleyan University Press, 1968.

Jones, Archer. *The Art of War in the Western World*, New York: Oxford University Press, 1989.

Kaplan, Robert D. *Imperial Grunts: The American Military on the Ground*, New York, NY: Random House, 2005.

Kearney, Thomas A. and Eliot A. Cohen. *Gulf War Airpower Survey: Summary Report*, Washington DC: U.S. Government Printing Office, 1993.

Keegan, John. *The Face of Battle*, London: Hutchinson, 1976.

.... *A History of Warfare*, Toronto, ON: Key Porter Books, 1993.

Kennedy, Paul. *The Rise and Fall of the Great Powers: Economic Change and Military Conflict from 1500 to 2000*, New York: Random House, 1987.

Kissinger, Henry. "The Nature of the National Dialogue on Foreign Policy", *Pacem in Terris III The Nixon-Kissinger Foreign Policy: Opportunities and Contradictions*, Fred Warner Neal and Mary Kersey Harvey, editors., Washington, DC, 1974.

.... *The White House Years*, Boston: Little, Brown and Co., 1979.

Kitchen, Martin. *A Military History of Germany from the Eighteenth Century to the Present Day*, London: Weidenfeld and Nicholson, 1975.

Kolko, G. *The Politics of War: The World and United States Foreign Policy, 1943-1945*, New York: Random House, 1968.

Kuhn, Thomas. *The Nature of Scientific Revolutions*, Chicago, IL: University of Chicago Press, 1996.

Lachester, F.W. *Aircraft in Warfare: The Dawn of the Fourth Arm*, London: Constable and Co., 1916.

Legault, Albert and Joel Sokolsky, editors, *The Soldier and the State in the Post Cold War Era*, Kingston, ON: Queen's Quarterly Press, 2002.

Leonhard, Robert. *The Art of Maneuver: Maneuver Warfare Theory and AirLand Battle*, Novato, CA: Presidio Press, 1991.

………*Fighting by Minutes*, Westport, CT: Praeger, 1994.

………*The Principles of War for the Information Age*, Novato, CA: Presidio Press, 1998.

Lider, Julian. *Military Theory: Concept, Structure, Problems*, Aldershot: Gower, 1983.

Lupfer, Timothy. *The Dynamics of Doctrine: The Changes in German Tactical Doctrine During the First World War*, Leavenworth Papers, No. 4., Leavenworth, KA: U.S. Army Command and Staff College, July 1981.

Luttwak, Edward. *The Political Uses of Sea Power*, Baltimore, MD: Johns Hopkins University Press, 1974.

Macdougall, P.L. *The Theory of War*, London: Longman, Brown, Green, Longman & Roberts, 1856.

MacGregor, Douglas. *Breaking the Phalanx*, Westport, CT: Praeger, 1997.

Macklin, W.H.S. *The Principles of War*, Canadian Army Training Pamphlet 5M-8-54 (M-7634-488), April 1948.

Macksey, Kenneth. *First Clash*, New York: Berkley Books, 1988.

……… *Panzer Division*, New York: Ballantine, 1968.

………*Guderian: Creator of Blitzkrieg*, New York: Stein and Day, 1976.

von Manstein, Erich. *Lost Victories*, Anthony G. Powell, translator, London: Methuen, 1994.

Marshall, S. L. A. *Men Against Fire: The Problem of Battle Command in Future War*, Gloucester, ME: Peter Smith, 1978.

Matloff, Maurice, general editor, *Army Historical Series: American Military History*, Washington, DC: U.S. Government Printing Office, 1985.

May, Ernest R. *Strange Victory: Hitler's Conquest of France*, New York: Hill and Wang, 2000.

……… and Richard E. Neustadt. *Thinking in Time: The Use of History for Decision Makers*, New York: New York Free Press, 1986.

McKercher, B.J.C. and M.A. Hennessy, editors, *The Operational Art: Developments in the Theories of War*, Westport, CT: Praeger, 1996.

McMurtry, John. *Understanding War*, Canadian Papers in Peace Studies, 1988, No. 2, Toronto, ON: University of Toronto Press, 1989.

McNeill, William H. *The Pursuit of Power; Technology, Armed Force, and Society since AD 1000*, Chicago, IL: University of Chicago Press, 1982.

von Mellenthin, F.W. *Panzer Battles: A Study of the Employment of Armour in the Second World War*, L.CcF. Turner, editor, H. Betzler translator, London: Futura Publications Ltd, 1979.

Melzer, Arthur M., Jerry Weinberger and M. Richard Zinman, editors, *History and the Idea of Progress*, Ithaca, NY: Cornell University Press, 1995.

Messenger, Charles. *The Blitzkrieg Story*, New York: Scribners, 1976.

.... *The Last Prussian: A Biography of Field Marshal Gerd von Rundstedt 1875-1953*, London: Brassey's, 1991.

Mets, David, R. *The Air Campaign: John Warden and the Classical Airpower Theorists*, Maxwell AFB, AL: Air University Press, 1999.

Midlarsky, Manus I., editor, *Handbook of War Studies*, Boston, MA: Unwin Hyman, 1989.

Miksche, F.O. *Blitzkrieg*, London: Faber & Faber, 1942.

.........*Atomic Weapons and Armies*, London: Faber & Faber, 1955.

Millett, Allan R. and Peter Maslowski. *For the Common Defense: A Military History of the United States of America*, New York: The Free Press, 1984.

Montross, Lynn. *War Through the Ages*, New York: Harper and Row, 1960.

Moses, John A. *The Politics of Illusion: The Fischer Controversy in German Historiography*, Brisbane, Australia: University of Queensland Press, 1975.

Murray, Williamson and Allan R. Millett, editors, *Military Innovation in the Interwar Period*, New York: Cambridge University Press, 1996.

National Defense University. *The Art and Practice of Military Strategy*, Washington, DC: 1984.

Oetting, Dirk W. *Auftragstaktik: Geschichte und Gegenwart einer Führungskonzeption*. Frankfurt am Main: Report Verlag, 1993.

Oman, Charles. *The Art of War in the Middle Ages*, London: Greenhill Books, 1991.

Palmer, R.R. *The World of the French Revolution*, New York: Harper, 1971.

Paret, Peter. *Clausewitz and the State: The Man, His Theories and His Times*, Princeton, NJ: Princeton University Press, 1976.

........., *Innovation and Reform in Warfare*, The Harmon Memorial Lectures in Military History, No. 8, United States Air Force Academy, 1966.

.... editor, with Gordon Craig and Felix Gilbert, *Makers of Modern Strategy from Machiavelli to the Nuclear Age*, Princeton, NJ: Princeton University Press, 1986.

Parker, Geoffrey. *The Military Revolution, Military Innovation and the Rise of the West, 1500-1800*, New York: Cambridge University Press, 1988.

Posen, Barry. *The Sources of Military Doctrine*, Ithaca, NY: Cornell University Press, 1984.

Preston, Richard A., Alex Roland and Sydney F. Wise. *Men in Arms: A History of Warfare and its Interrelationships with Western Society*, Fort Worth, TX: Harcourt Brace Jovanovich 1991.

Pruitt, Dean G. and Richard C Snyder, editors, *Theory and Research on the Causes of War*, Englewood Cliffs, NJ: Prentice-Hall, 1969.

Reardon, Carol. *Soldiers and Scholars: The U.S. Army and the Uses of Military History, 1865-1920*, Lawrence, KS: University of Kansas Press, 1990.

Reid, Brian Holden. *J.F.C. Fuller: Military Thinker*, Basingstoke, UK: Macmillan, 1987.

Reynolds, Clark. *Command of the Sea: The History and Strategy of Maritime Empires*, New York: William Morrow and Co., 1974.

Richmond, Herbert. *National Defence: The Navy*, London: William Hodge & Company, 1937.

Ritter, Gerhard. *The Sword and the Sceptre: The Problem of Militarism in Germany*, 4 Vols., Coral Gables, FL: University of Miami Press, 1969.

Roberts, Michael. *The Military Revolution, 1560-1660*, Inaugural Lecture. Belfast: Queen's University of Belfast Press, January 1955.

Theodore Ropp. "Continental Naval Theories," *Makers of Modern Strategy from Machiavelli to Hitler*, Edward Mead Earle, editor, (Princeton, NJ: Princeton University Press, 1973).

Rosinsky, Herbert. *The German Army*, London: Pall Mall Press, 1966.

Rothenberg, Gunther. *The Art of Warfare in the Age of Napoleon*, Bloomington, IN: Indiana University Press, 1990.

Scales, Robert H. *Future Warfare*, Carlisle, PA: U.S. Army War College, 1999.

Schreiber, Shane. *Shock Army of the British Empire: The Canadian Corps in the Last 100 Days of the Great War*, Westport, CT: Greenwood Press, 1997.

Schroeder, Paul. *The Transformation of European Politics 1763-1848*, Oxford: Oxford University Press, 1994.

von Seeckt, Hans. *Thoughts of a Soldier*, Gilbert Waterhouse, translator, London: Ernest Benn, 1930.

de Seversky, Alexander. *Victory Through Air Power*, Garden City, NY: Garden City Publishing, 1943.

Shanahan, Walter. *Prussian Military Reforms 1786-1813*, New York: Columbia University Press, 1945.

Shaw, Martin. *The Dialectics of War: An Essay in the Social Theory of War and Peace*, London: Pluto, 1988.

Showalter, David. *Railroads and Rifles: Soldiers, Technology, and the Unification of Germany*, Hamdon, CT: Archon Books, 1975.

Speelman, Patrick. *Henry Lloyd and the Military Enlightenment of Eighteenth Century Europe*, Westport, CT: Greenwood Press, 2002.

Stoessinger, John. *Why Nations Go to War*, New York: St Martin's Press, 1990.

Stokesbury, James L. *A Short History of Air Power*, New York: William Morrow & Co., 1986.

Strachan, Hew. *European Armies and the Conduct of War*, London: Unwin Hyman, 1983.

Strategic Studies Institute. *The Principles of War in the 21st Century: Strategic Considerations*, Carlisle, PA: U.S. Army War College, 1995.

Sullivan, Gordon R., General, and Colonel James M. Dubik. *Envisioning Future*

Warfare, Fort Leavenworth, KA: U.S. Army Command and General Staff College, 1995.
Summers, Harry. *On Strategy: The Vietnam War in Context*, Carlisle Barracks, PA: U.S. Army War College, 1983.
Taylor, A.J.P. *The Origins of the Second World War*, Greenwich: Fawcett, 1961.
Toffler, Alvin. *Future Shock*, New York: Random House, 1970.
.... *The Third Wave*, New York: William Morrow and Co., 1980.
Trainor, Bernard E. and Michael R. Gordon. *The Generals' War: The Inside Story of the Conflict in The Gulf*, Boston, MA: Little, Brown and Co., 1995.
Tuchman, Barbara. *The Guns of August*, New York: Macmillan, 1962.
.... *Practising History: Selected Essays*, Westminster, MD: Ballantine Books, 1982.
.... *The March of Folly: From Troy to Vietnam*, New York: Alfred A. Knopf, 1984.
Vigor, P.H. *Soviet Blitzkrieg Theory*, New York: St Martin's, 1983.
Wallach, Jehuda. *The Dogma of the Battle of Annihilation: The Theories of Clausewitz and Schlieffen and Their Impact on the German Conduct of Two World Wars*, Westport, CT: Praeger, 1986.
Warden, John A. *The Air Campaign*, Washington, DC: Brassey's, 1991.
Weigley, Russell. *History of the United States Army* enlarged ed., Bloomington, IN: Indiana University Press, 1984.
.... *The American Way of War: A History of the United States Military Strategy and Policy*, Bloomington, IN: Indiana University Press, 1984.
West Point Department of Military Art and Engineering. *Jomini, Clausewitz and Schlieffen*, 1954.
White, Charles Edward. *The Enlightened Soldier: Scharnhorst and the Militärische Gesellschaft in Berlin, 1801-1805*, Westport, CT: Praeger, 1989.
Wilkinson, Spenser. *The Brain of an Army, The German General Staff*, London: Macmillan, 1890.
.... "Strategy in the Navy", *The Morning Post*, London, 3 August 1909.
Williams, Harry T. *Americans at War: The Development of the American Military System*, Baton Rouge, LA: Louisiana State University Press, 1960.
.... *Lincoln and his Generals*, New York: Alfred A. Knopf, 1952.
Wylie, J. C. *Military Strategy: A General Theory of Power Control*, Annapolis, MD: U.S. Naval Institute, 1967.
Zook, David H., and Robin Higham. *A Short History of Warfare*, New York: Twayne Publishing, 1966.

Articles and Chapters

Adams, Thomas, K. "Future Warfare and the Decline of Human Decisionmaking",

Parameters, (Winter 2001-2002), 57-71.

Bashista, Ronald J. "Auftragstaktik It's More Than Just a Word," *U.S. Army Armor*, (November-December 1994).

Baucom, Donald R. "Modern Warfare: Paradigm Crisis?" *Air University Review*, (March-April 1984), 2-3.

Bayerlein, Fritz. "With the Panzers in Russia 1941 & 43," *Marine Corps Gazette*, (December 1954).

Belote, Howard D. "Warden and the Air Corps Tactical School: What Goes Around Comes Around," *Airpower Journal*, (Fall 1999).

Bolia, Robert S. "Overreliance on Technology in Warfare: The Yom Kippur War as a Case Study", *Parameters* (Summer 2004), 46-56.

Brown B.G. "Manoeuvre Warfare Roadmap, Part I Trends and Implications", *Marine Corps Gazette*, (April 1982), 42-47.

Carlson, Verner R. "Portrait of a German General Staff Officer" *Military Review* (April 1990), 69-81.

Dickson, Keith D. "War in (Another) New Context: Post Modernism," *The Journal of Conflict Studies* (Winter 2004), 78-91.

Eriksson, Anders E. "Information Warfare: Hype or Reality?" *The Nonproliferation Review* (Spring-Summer 1999).

van Creveld, Martin. "On Learning From the Wehrmacht and Other Things" *Military Review* (January 1988), 62-71.

Crowl, Philip A. "Alfred Thayer Mahan: The Naval Historian," *Makers of Modern Strategy: From Machiavelli to the Nuclear Age*, Peter Paret editor, Princeton: Princeton University Press, 1986.

Dahl, Eric J. "Net-Centric Before its Time: The *Jeune École* and its Lessons for Today" *Naval War College Review* (Autumn 2005), 110-135.

De Vries P.T. "Manoeuvre and the Operational Level of War", *Military Review* (February 1983), 13-33.

Dunlap, Charles J., Jr. "21st-Century Land Warfare: Four Dangerous Myths" *Parameters* (Autumn 1997), 27-37.

.... "Technology: Recomplicating Moral Life for the Nation's Defenders", *Parameters* (Autumn 1999), 24-53.

.... "Moltke and the German Military Tradition: His Theories and Legacies" *Parameters* (Spring 1996).

.... "War, Politics, and RMA-The Legacy of Clausewitz," *Joint Forces Quarterly* (Winter 1995-96).

Fleming, Bruce. "Can Reading Clausewitz Save Us from Future Mistakes?" *Parameters* (Spring 2004), 62-76.

Forsyth, Michael. "Finesse: A Short Theory of War," *Military Review* (July-August 2004), 17-19.

Förster, Stig. "Facing 'People's War': Moltke the Elder and Germany's Military

Options after 1871" *The Journal of Strategic Studies* (June 1987), 209-230.

Franz W.P. "Manoeuvre: The Dynamic Element of Combat", *Military Review* (May 1983), 2-12.

Fukuyama, Francis. "The End of History," *National Interest* (Summer 1989), 3-18.

Gentry John A. "Doomed to Fail: America's Blind Faith in Military Technology", *Parameters* (Winter 2002-03), 88-103.

Gilbert, Felix. "Machiavelli - The Renaissance of the Art of War." *Makers of Modern Strategy: from Machiavelli to The Nuclear Age*. Peter Paret, editor, Princeton: Princeton University Press, 1986.

Gleick, James. "Exploring the Labyrinth of the Mind." *N Y Times Magazine* (21 August 1983).

Goulding, Vincent J. Jr. "From Chancellorsville to Kosovo, Forgetting the Art of War", *Parameters* (Summer 2000), 4-18.

Gray Colin S. "How Has War Changed Since the End of the Cold War?" *Parameters* (Spring 2005), 14-26.

Handel, Michael I. "Corbett, Clausewitz, and Sun Tzu" *Naval War College Review*: Vol. 53 (2000): No. 4,

.... "Clausewitz in the Age of Technology," *Journal of Strategic Studies* (June-September 1986).

Hanson, Victor Davis. "The Western Way of War" *Australian Army Journal*, (Winter 2004), 157-164.

Henry, Ryan and C. Edward Peartree. "Military Theory and Information Warfare", *Parameters* (Autumn 1998), 121-135.

Hill, J.R. "Accelerator and Brake: The Impact of Technology on Naval operations, 1855-1905," *Journal for Maritime Research* (December, 1999).

Holborn, Hajo. "The Prusso-German School - Moltke and the Rise of the General Staff." *Makers of Modern Strategy: from Machiavelli to The Nuclear Age*, Peter Paret, editor, Princeton: Princeton University Press, 1986.

Hooker, R. D. "Beyond *Vom Kriege*: The Character and Conduct of Modern War", *Parameters* (Summer 2005), 4-17.

Hope, Ian. "Misunderstanding Mars and Minerva: The Canadian Army's Failure to Define an Operational Doctrine", *Army Doctrine and Training Bulletin* (Winter 2001-2002), 16-35.

Huntington, Samuel P. "Clash of Civilizations," *Foreign Affairs* (Summer 1993), 22-49.

Johnston, Paul. "Doctrine Is Not Enough: The Effect of Doctrine on the Behavior of Armies", *Parameters* (Autumn 2000), 30-39.

Kagan, Frederick. "Army Doctrine and Modern War: Notes Toward a New Edition of FM 100-5," *Parameters* (Spring 1997), 134-51.

Kazarin, P.S. "The Nature of War as a Scientific Category", *Military Thought*

(2002).
Michael Kelly. "The Air-Power Revolution", *The Atlantic Monthly* (April 2002), 18-20
.... "Slow Squeeze." *The Atlantic Monthly* (May 2002), 20-22.
.... "American War Air-Power." *The Atlantic Monthly* (June 2002), 10-11.
Kitchen, Martin. "The Traditions of German Strategic Thought," *The International History Review* (April 1979), 163-190.
Krause, Michael D. "Moltke and the Origins of Operational Art" *Military Review* (September 1990), 28-44.
Lind William S. "Defining Manoeuvre Warfare for the Marine Corps", *Marine Corps Gazette* (March 1980), 55-58.
.... Keith Nightengale, John F. Schmitt, Joseph W. Sutton, and Gary I. Wilson "The Changing Face of War: Into the Fourth Generation" *Marine Corps Gazette* (October 1989), 22-26.
.... John F. Schmitt and Gary I. Wilson. "Fourth Generation Warfare: Another Look," *Marine Corps Gazette* (December 1994), 34-37.
Linn, Brian M. "*The American Way of War* Revisited", *Journal of Military History* (April 2002), 501-533.
Lwin, Michael R. "General Tzu's Army: OPFOR of the Future." *Joint Forces Quarterly* (Spring 1997), 44-49.
Morton, Louis. "National Policy and Military Strategy", *Virginia Quarterly Review*, (Winter 1960).
Murdock, Paul. "Principles of War on the Network-Centric Battlefield: Mass and Economy of Force" *Parameters* (Spring 2002), 86-95.
Murray, Williamson. "The Army's Advanced Strategic Art Program", *Parameters* (Winter 2000-01), 31-9.
.... "Thinking About Innovation, *Naval War College Review* (Spring 2001), 119-130.
.... "War, Theory, Clausewitz and Thucydides: The Game May Change But the Rules Remain", *Marine Corps Gazette* (January 1997), 62-69.
Parker, Geoffrey. "The 'Military Revolution,' 1560-1660 - A Myth?" *Journal of Modern History* (June 1976), 195-214.
Reid, Brian Holden. "J.F.C. Fuller's theory of mechanized warfare", *Journal of Strategic Studies* (1978).
Rippe S.T. "Leadership, Firepower and Manoeuvre: The British and Germans, *Military Review* (October 1985), 30-36.
Ropp, Theodore. "Continental Doctrines of Sea Power," Edward Meade Earle, editor, *Makers of Modern Strategy: Military Thought from Machiavelli to Hitler*, Princeton: Princeton University Press, 1941.
Rothenberg, Gunther E. "Moltke, Schlieffen, and the Doctrine of Strategic Envelopment," *Makers of Modern Strategy: from Machiavelli to The Nuclear Age*,

Peter Paret editor, Princeton: Princeton University Press, 1986

Schneider, James J. "Black Lights: Chaos, Complexity, and the Promise of Information Warfare." *Joint Forces Quarterly* (Spring 1997), 21-28.

Simpkin R. "Manoeuvre Theory and the Small Army", *British Army Review* (December 1984).

Stone, J.C. "The Canadian Army's Principles of War in the Future: Are they Relevant?", *The Army Doctrine and Training Bulletin* (Spring 2000).

Waldron, Arthur. "China's Military Classics," *Joint Forces Quarterly* (Spring 1994), 114-117.

Monographs and Lectures

Advanced Information Booklet, *Robert Aumann's and Thomas Schelling's Contributions to Game Theory: Analyses of Conflict and Cooperation*, Royal Swedish Academy of Sciences, Bank of Sweden Prize in Economic Sciences in Memory of Alfred Nobel, Stockholm, 10 October 2005.

Buchan, Glenn C. *Future Directions in Warfare: Good and Bad Analysis, Dubious Rhetoric, and the "Fog of Peace"* Santa Monica CA: Rand Corporation, 2003.

Crowl, Philip A. *The Strategist's Short Catechism: Six Questions Without Answers*, The Harmon Memorial Lectures in Military History, No. 20, United States Air Force Academy, 1977.

Dick, C. J. *Conflict in a Changing World: Looking Two Decades Forward*, Conflict Studies Research Centre, Directorate General Development and Doctrine, Royal Military Academy Sandhurst, Camberley Surrey, Feb 2002

Dubik, James M. *Has Warfare Changed? Sorting Apples from Oranges*, Landpower Essay No. 02-3, Association of the United States Army Institute of Land Warfare July 2002.

Durham, Susan E. "Chaos Theory for The Practical Military Mind", Research Paper presented to United States Air Command and Staff College, March 1997.

Echevarria, Antulio J. II. "Toward an American Way of War", Monograph, Strategic Studies Institute, U.S. Army War College, Carlisle Pennsylvania, March 2004.

.... "Globalization and the Nature of War", Monograph, Strategic Studies Institute, U.S. Army War College, Carlisle Pennsylvania, March 2003.

English, Allan. "The Operational Art: Theory, Practice, and Implications for the Future", Monograph for Comment, Canadian Forces Command and Staff College, Toronto Ontario, January 2003.

Faber, Peter. "Strategy: Its Intellectual Roots and Its Current Status", excerpted from "The Evolution of Airpower Theory in the United States – From World War I to John Warden's The Air Campaign," *Asymmetric Warfare*, John Andreas Olsen editor, Norwegian Air Force Academy Militærteortisk Skriftserie No. 4,

2002.

Fadok, David S. "John Boyd and John Warden: Air Power's Quest for Strategic Paralysis," Monograph School of Advance Airpower Studies, U.S. Air Force Air University, Maxwell Air Force Base, Alabama, February 1995.

Feldman, Shai. *Technology and Strategy: Future Trends*, Jaffee Center for Strategic Studies Conference Summary, Boulder. CO: Praeger Press, 1989.

The Future of Warfare: Issues from the 1999 Army After Next Study Cycle, U.S. Army TRADOC publication, 2000.

Hall, Mike. "Effects Based Operations – A Primer", Monograph written for Joint Forces Command, Suffolk Virginia 18 Apr 2003.

Howard, Michael. *Strategy and Policy in Twentieth-Century Warfare*, Harmon Memorial Lectures in History, No. 9, United States Air Force Academy 5 May 1967.

Johnsen, William T., Douglas V. Johnson II, James O. Kievit, Douglas C. Lovelace, Jr., and Steven Metz. "The Principles of War in the 21st Century: Strategic Considerations," Monograph from Strategic Studies Institute, U.S. Army War College, Carlisle, PA, 1 August 1995.

Jarymowycz, Roman. "The Quest for Operational Maneuver in the Normandy Campaign: Simonds and Montgomery Attempt the Armoured Breakout," PhD diss., McGill University, 1997.

Olivier, David Harold. "*Staatskaperei*: The German Navy and Commerce Warfare: 1856-1888," PhD diss., University of Saskatchewan, 2001.

Oliviero, Charles S. "The Early Development of Auftragstaktik," MA diss., Royal Military College of Canada, 1995.

Paret, Peter. *Innovation and Reform in Warfare*, The Harmon Memorial Lectures in Military History, No. 8, United States Air Force Academy, 1966.

Pellegrini, Robert P. "The Links Between Science, Philosophy, and Military Theory", Monograph School of Advanced Airpower Studies, Maxwell Air Force Base, Alabama, Aug 1997.

Ropp, Theodore. *The Historical Development of Contemporary Strategy*, The Harmon Memorial Lectures in Military History, No. 12, United States Air Force Academy, 1970.

Sharpe, S.J. *Principles of War for Canada in the 21st Century* A Research Paper presented to Canadian Forces Command and Staff College, June 2003.

Storr, J.P. *The Command of British Land Forces in Iraq, March to May 2003*. Directorate General of Development and Doctrine, British Army, Spring 2004, Wiltshire UK.

.... "The Nature of Military Thought", PhD diss., Cranfield University, UK, 2002.

Studer Juerg, "Are There Five Rings or a Loop in Fourth Generation Warfare? A Study on the Application of Warden's or Boyd's Theories in 4GW." Monograph for U.S. Air Command and Staff College, Maxwell Air Force Base, Alabama, April 2005.

Strategy and Defense Planning for the 21st Century, Santa Monica: Rand Corporation, 1997.

U.S. Army War College Guide to Strategy, Cerami, Joseph R. and James F. Holcomb, Jr. editors, February 2001.

Williams, Harry T. *The Military Leadership of the North and South*, The Harmon Memorial Lectures in Military History, Number 2, (Colorado Springs, CO: United States Air Force Academy, 1960).

Internet

Battlefield of the Future: 21st Century Warfare Issues. U.S.A.F. Air University, http://www.airpower.maxwell.af.mil/airchronicles/battle/bftoc.html.

Butner, Ernest. *The Art of War (Machiavelli, Vauban, and Frederick)*. http://civilwarhome.com/artofwar.htm

Betts, Richard K., *Should Strategic Studies Survive?* World Politics 50.1 (1997) 7-33. http://muse.jhu.edu/journals/world_politics/v050/50.1betts.html

von Clausewitz, Carl, *Vom Kriege*, (Originally published by Dümmlers Verlag, Berlin, 1832) http://www.clausewitz.com/CWZHOME/CWZBASE.htm

Cooper, Gary T. *Naval Maneuver Warfare*. Available from http://www.globalsecurity.org/military/library/report/1995/CG.htm

Falk, Dan. *The National Post*. http://www.nationalpost.com/search/story.html?f=/stories/20020123/1212674.html&qs=Dan%20Falk

Fayette, Daniel F. "Effects-Based Operations: Application of new concepts, tactics, and software tools support the Air Force vision for effects-based operations," U.S. Air Force Research Laboratory, Information Directorate, Information Technology Division, Dynamic Command and Control Branch, Rome NY. http://www.afrlhorizons.com/Briefs/June01/IF00015.html.

Garden, Timothy. *Writings*. Available from http://www.tgarden.demon.co.uk.

Goh Teck Seng. "Airpower as the "Fulcrum" of Modern Conventional Warfare?" http://www.mindef.gov.sg/safti/pointer/back/journals/1998/Vol24_4/4.htm

Hinen, Col Anthony L. "Kosovo: 'The Limits of Air Power II': War Can Be Won With Airpower Alone!," 16 May 2002 *Air & Space Power Chronicles*. http://www.airpower.maxwell.af.mil/airchronicles/cc/hinen.html.

Keller, Bill. "The Fighting Next Time" *New York Times Magazine*. http://www.nytimes.com/2002/03/10/magazine/10MILITARY.html.

Lambe, Patricke. "The Perils of Knowledge-Based Warfare," April 2003. http://www.destinationkm.com/articles/default.asp?ArticleID=1043.

Lind, William S. *Understanding Fourth Generation War*. http://www.military.com/

NewContent/0,13190,Lind_121903,00.html

Murray, Williamson. "Military Culture Does Matter." http://www.fpri.org/fpriwire/0702.199901.murray.militaryculturedoesmatter.html

New, Larry D. "Clausewitz's Theory: On War and Its Application Today" *Airpower Journal* Fall 96. http://www.airpower.maxwell.af.mil/airchronicles/apj/fal96.html

Project MUSE, Johns Hopkins University, http://muse.jhu.edu/proj_descrip/gen_intro.html

Watts, Barry D. *Clausewitzian Friction and Future War*, McNair Paper No. 52, October 1996. http://www.ndu.edu-ndu-inss-macnair-mcnair52-m52cont.html

Weeks, Major Michael R. "Chaos, Complexity and Conflict," 16 July 2001 *Air & Space Power Chronicles*. http://www.airpower.maxwell.af.mil/airchronicles/cc/Weeks.html

Yarger, Richard. "Towards A Theory of Strategy: Art Lykke and the Army War College Strategy Model" U.S. Air War College Internet Homepage. http://dde.carlisle.army.mil/authors/stratpap.htm

GENERAL INDEX

air power, 39, 45, 46, 51n, 71, 72, 84, 87n, 88n, 92, 98, 104, 109n, 111-127, 130-137, 140, 142-144, 149, 150, 169, 180, 181, 184n, 189, 190, 192-194.
aufklärung, 41, 43, 45, 63, 66.
auftragstaktik, 36, 40, 41-43, 50, 63, 65, 71, 80, 154, 171.
bombing, 112, 113, 120, 123, 126, 134, 181.
    conventional, 125.
    precision, 72, 123, 133.
    strategic, 46, 116, 117, 119, 121, 122, 124, 135, 194.
blitzkrieg, 37, 70, 81-83, 135, 143.
centre of gravity, 7, 49, 67, 112, 134.
Cold War, 71, 92, 111, 124-127, 130-132, 135, 164n, 168, 173-175, 179, 180, 185n, 191, 193, 195, 196.
conceptual grafting, 38, 171.
culture, 7, 9, 10, 13-17, 21, 33, 35, 37, 39, 40, 41, 44, 45, 50, 51, 63, 139, 142, 152, 153, 171, 172, 180, 181, 187, 188, 195, 19.
doctrine, 7, 12, 26, 27, 29, 30, 33, 36-39, 42-44, 71, 72, 81, 92, 93, 101, 103, 117, 119, 121, 122, 134, 135, 144, 147, 151, 152, 154, 157, 159, 169, 171, 172, 177, 188, 190, 193.
Effects Based Operations (EBO),119, 120, 134, 139, 147, 163, 167, 183, 193, 194.
enlightenment, 41, 46-48, 51, 58, 63, 87n, 145, 162.
4th Generation Warfare (4GW), 46, 59, 139, 140, 147, 148, 163, 167, 183.
game Theory, 128-130.
general staff, 23, 38, 39, 44, 51n, 63, 71, 79-82, 88n, 114, 117, 127, 147, 171.
guerre de course , 93, 96, 98, 107, 190
guerre d'escadre, 93, 98, 106.
Jeune École, 2, 93-97, 107, 108n, 130, 182, 190, 193.

manoeuvre, 7, 42, 70, 71, 73, 76, 81, 114, 119, 134, 139, 141, 153-155, 159, 190, 191.
mission analysis, 39, 42, 43.
Mutually Assured Destruction (MAD), 128.
NATO, 7-9, 11, 26, 30, 39-43, 50, 72, 102, 118, 119, 126, 132-135, 151, 154, 157, 159, 162, 174, 188.
nuclear, 2, 20, 71, 87n, 92, 103, 104, 106, 108n, 124-132, 135, 137n, 140, 142, 147, 152, 167, 169, 174, 175, 181, 203.
ontology, 14, 15, 187, 204.
OODA Loop, 155-157.
operational art (Level), 7, 13, 25, 41, 49, 51n, 78, 81-83, 101, 102, 105, 127, 136, 137n, 144, 150, 158, 164n, 165n, 171, 173, 189.
Professional Military Education (PME), 24, 65.
Prussian, 11, 13, 26, 40, 41, 44, 61-63, 69, 71, 76, 77, 79, 80, 95, 96, 113, 160, 161, 171, 189, 190.
RAND Corporation, 125, 127, 130, 132, 172, 173, 184n.
Renaissance, 37, 55, 59, 161, 195.
revolution, 39, 48.
French, 26, 40, 56, 63, 78, 189, 190.
industrial, 46, 47, 64.
military, 62, 72, 73, 87, 88n, 89n, 104, 139, 141-150, 152, 154, 162, 164n, 175, 178, 189.
School of Advanced Military Studies, 71, 72.
schwerpunkt, 27, 171.
sea power, 91-100, 106, 108n, 109n, 135, 149, 191.
Soviet, 13, 70, 83, 84, 103, 104, 121, 124, 125, 131, 142, 147, 162, 174.
strategy, 7, 8, 11-14, 16, 19, 22, 24, 28, 29, 36, 40, 41, 44, 46, 50, 51n, 59, 60, 61, 66, 69, 72, 77, 80-85, 87n, 88n, 89n, 92-97, 100, 102-104, 106, 107, 108n, 122, 124-133, 135, 137n, 143, 152, 162, 173, 177, 180, 183, 184, 184n, 187-190
submarine, 93, 103-105.
synchronization, 7, 154.
tactics (tactical) , 11-13, 24-26, 36-38, 40, 41, 44, 46, 49-51, 55, 60, 61, 63, 66, 69, 70, 73-86, 88n, 93, 94, 97, 100-102, 104-106, 111, 112, 116, 117, 120-122, 124-127, 134, 135, 139, 143, 144, 150-152, 156, 162, 171, 173, 176, 183, 188, 192, 195, 197.
taxonomy, 11, 14, 187.
Wehrmacht, 37, 49, 81, 82, 144, 153.

INDEX OF PERSONALITIES

Alexander the Great, 19, 72, 85, 139, 156, 172, 189.
Ager, John, 85, 87n.
Aristotle, 19, 203.
Aube, Hyacinthe, 93, 96, 130.
Aumann, Robert J., 128-130.
Bismarck, Otto von, 80, 95.
Boyd, John, 2, 59, 111, 118, 119, 148, 155-157, 159.
Brodie, Bernard, 28, 29, 130, 132, 167.
Bülow, Heinrich von, 62, 63, 189.
Caesar, Julius, 2, 72, 160, 189.
Chaliand, Gérard,19, 20, 31n.
Citino, Robert, 37.
Clausewitz, Carl von, 1, 6, 7, 8, 10, 11, 17, 20-24, 26-29, 31n, 34, 41, 43, 45, 49, 51n, 59, 61-67, 69, 70, 72, 79-81, 84-87, 99-101, 105, 108n, 109n, 113, 116, 131, 149, 152, 154, 156, 158, 161, 173, 176, 177, 180, 185, 189-195, 197, 198, 202.
Corbett, Julian,2, 23, 91, 93, 96, 97, 100-107, 108n, 109n,190, 193.
Colomb, John and Philip,23, 93, 96, 97, 102, 108n.
Darwin, Charles, 47, 99, 158, 163.
De Gaulle, Charles, 19.
Delbrück, Hans, 38, 39.
Douhet, Giulio, 2, 23, 24, 26, 39, 46, 111-117, 119-126, 131-135, 136n, 180, 189-191, 193.
Dupuy, Trevor, 37, 81.
Echevarria, Antulio, 27, 83, 89n, 134.
Fischer, Fritz, 38, 39.
Frederick II (The Great), 23, 39, 40, 44, 49, 76, 77, 79, 81, 87, 88n, 89n, 160, 189.

Freedman, Lawrence, 126, 137n.
Fuller, J.F.C, 28, 59, 69-71, 81, 88, 115, 136n, 151, 160, 161, 190, 197.
Galileo, 47, 66.
Gat, Azar, 51n, 136n, 202.
Gneisenau, August von, 23, 43, 62.
Goltz, Colmar von der, 6, 54.
Guderian, Heinz, 70, 81, 155, 156.
Gustavus Adolphus, 49, 73-76, 87n, 88n, 156, 171, 189.
Halleck, Henry Wager, 26, 67-69, 72, 88n.
Handel, Michael, 8, 27, 38, 156, 161.
Hegel, Georg Wilhelm, 8, 27, 38, 156, 161.
Ho Chi Minh, 53.
Howard, Michael, 29, 37, 49, 51n, 88n, 180, 202.
Huntington, Samuel, 30, 31, 179, 180, 194.
Jomini, Henri de, 6, 7, 17, 20, 21, 22-24, 26-29, 31, 34, 61, 62, 64, 65, 67-69, 79, 85-88, 98, 99, 105, 158, 160, 189-193, 198.
Keegan, John, 37, 202.
Lawrence, T.E., 54, 146.
Lincoln, Abraham, 11, 49, 146.
Liddell Hart, Basil, 28, 36, 59, 69, 70, 81, 88n, 136n, 190, 194, 201.
Lloyd, Henry, 60-62, 64, 68, 75, 86, 87n, 189, 191,192.
Machiavelli, Niccolò, 1, 20, 21, 23, 28, 29, 45, 57-61, 72-74, 86, 87, 108n, 152, 161, 180, 189, 191, 197.
Mahan, Alfred Thayer, 2, 23, 24, 26, 91-93, 96-107, 108n, 159, 189-191, 193.
Mahan, Dennis Hart, 67-69.
Manstein, Erich von, 35, 84, 87, 155.
Mao Zedong, 9, 53, 183.
Marx, Karl, 6, 180.
Matrix, 12-15, 17, 19, 24, 25, 33, 35, 50, 53-55, 85, 129, 139, 154, 181, 183, 187, 188, 191, 192, 196, 198, 203, 204.
McAndrew, William, 43, 171.
Mitchel, William (Billy), 2, 111-113, 117, 118, 121, 132, 134, 135, 137.
Moltke, Helmuth von (Elder), 23, 28, 43, 44, 79-82, 89n, 95, 191.
Montecuccoli, Raimondo, 49, 59, 60, 61, 87n, 189.
Murray, Williamson, 22, 31n, 37, 45, 51n, 141, 164n, 178, 179, 185n, 196, 197, 199n.
Musashi, Myamoto, 5, 59, 69, 87n, 189.
Napoleon, 7, 13, 24-29, 40, 41, 48, 49, 51, 59, 62-64, 66, 68, 73, 76-79, 83, 86, 89n, 98, 101, 150-152, 156, 160, 183, 189, 192.
Newton, Isaac, 47, 48, 50, 61, 66, 67, 162, 188, 202, 203.
Parker, Geoffrey, 143.
Patton, George S., 23, 84, 136.

INDEX

Pellegrini, Robert, P., 51n, 88n.
du Picq, Charles Ardent, 24, 197.
Plato, 13, 14, 17, 23, 158, 188, 202.
Richmond, Herbert, 2, 92, 102, 108n.
Rommel, Erwin, 25, 84, 114, 127.
Scharnhorst, David Gerhardt von, 23, 41, 43, 62, 63, 65, 77, 88n, 189, 190.
Schelling, Thomas C., 127, 128, 130.
Schlieffen, Alfred von, 79, 80-82.
Seeckt, Hans von, 23, 82, 159.
Sun Tzu, 5-7, 10, 17, 20-22, 24, 26, 28, 29, 34, 45, 53, 56-59, 65, 69, 70, 85-87, 105, 108n, 109n, 148, 152, 161, 163, 165n, 177, 191, 192, 194, 197, 198.
Tirpitz, Alfred von, 93, 107.
Trenchard, Hugh, 2, 116, 117, 135.
Tukhachevsky, Mikhail, 54, 70.
Vegetius, Flavius Renatus, 57, 59, 60, 72, 87, 93, 161, 168, 189.
Warden, John, 2, 59, 87n, 111, 112, 118-120, 134-136.
Wass de Czege, 71, 72, 88n.
Weigley, Russel, 27, 69, 88n, 160.
Wilkinson, Spenser, 70, 101.

DOUBLE✠DAGGER

Double Dagger Books is Canada's newest military-focused publisher. Conflict and warfare have shaped human history since before we began to record it. The earliest stories that we know of, passed on as oral tradition, speak of war, and more importantly, the essential elements of the human condition that are revealed under its pressure. We are dedicated to publishing material that, while rooted in conflict, transcend the idea of "war" as merely a genre. Fiction, non-fiction, and stuff that defies categorization, we want to read it all.

Because if you want peace, study war.

www.doubledagger.ca

PRAXIS TACTICUM

THE ART, SCIENCE AND PRACTICE OF MILITARY TACTICS

COLONEL CHARLES S. OLIVIERO

ALSO BY COLONEL CHARLES S. OLIVIERO

PRAXIS TACTICUM
The Art, Science and Practice of Military Tactics

"Praxis Tacticum" will help young leaders learn and master modern combat team tactics…It's a fascinating series of intellectual and practical exercises which will help those leading fast moving and hard-hitting troops in combat, a unique blend of both the science and art of war.
Lieutenant-General (ret'd) The Hon. Andrew Leslie, PC, CMM, MSC, MSM, CD, MA, PhD

Chuck brings the discussion on tactics to the 'centre of gravity' between operational and theoretical perspectives. Praxis Tacticum is for professionals, people interested in tactics and the general reader of history.
Major-General (ret'd) David Fraser, CMM, MSC, MSM, CD

Pundits the world over have long predicted the end of conventional warfare but for the foreseeable future, it is here to stay. Counter insurgency, guerrilla warfare, terrorism, peace enforcement, policing duties. All of these forms, like conventional warfare, are as old as mankind. Modern militaries claim that they are professional bodies, responsible to teach, control and discipline their members. But at least one aspect of this claim is poorly executed: tactics are not taught to junior leaders, which is why this practical guide is essential reading for all junior leaders, officers and NCOs alike.

There is an old military adage that there is no teacher like the enemy. True; but the wise commander will prepare to meet that enemy and become the teacher and not the student.

ABOUT THE AUTHOR

Colonel (Retired) Chuck Oliviero, CD, PhD is an internationally recognized expert in simulation supported training and has twice been the Keynote Speaker at international training conferences and fora. He has over four decades' experience as an educator and trainer. For two decades, he was responsible for designing, developing and delivering some of the most complex collective training events ever conducted in synthetic environments for military, government and corporate entities. Colonel Oliviero served more than 30 years in the Canadian Forces, retiring as a Colonel. His career included command of Canada's then only tank regiment, The 8th Canadian Hussars (Princess Louise's), establishing Canada's Arms Control Verification Unit and being both an instructor and the Chief of Staff of the Canadian Army Command and Staff College. He is a graduate of the Royal Military College of Canada, holds a BA (Hon) in History, an MA and a PhD in War Studies. He is also a graduate of the two-year German War College (Führungsakademie der Bundeswehr) course. His last military duty was as Special Advisor to the Commander Canadian Army. In 2011 the Minister of National Defence appointed him as Honorary Lieutenant Colonel of the Queen's York Rangers (1st Americans). For more than a decade, he was an Adjunct Professor of history and strategy at both the Royal Military College and Norwich University in Vermont, USA. He is married and has two sons, both of whom are serving officers in the Canadian Armed Forces.

Made in the USA
Las Vegas, NV
30 October 2023

79897700R00138